AF845103

BIRMINGHAM UNIVERSITY LIBRARY

THIS BOOK MUST BE RETURNED IMMEDIATELY IF RECALLED FOR THE USE OF ANOTHER BORROWER, OTHERWISE ON OR BEFORE THE LATEST DATE BELOW

NEW BOOK

Optoacoustic Spectroscopy and Detection

Contributors

P. C. Claspy
C. Forbes Dewey, Jr.
Jerry A. Gelbwachs
P. L. Kelley
Lloyd B. Kreuzer
M. B. Robin
Allan Rosencwaig
John D. Stettler
Norman M. Witriol

OPTOACOUSTIC SPECTROSCOPY AND DETECTION

Edited by

YOH-HAN PAO

Department of Electrical Engineering
and Applied Physics
School of Engineering
Case Western Reserve University
Cleveland, Ohio

ACADEMIC PRESS New York San Francisco London 1977

A Subsidiary of Harcourt Brace Jovanovich, Publishers

1142916

BIRMINGHAM UNIVERSITY LIBRARY

COPYRIGHT © 1977, BY ACADEMIC PRESS, INC.
ALL RIGHTS RESERVED.
NO PART OF THIS PUBLICATION MAY BE REPRODUCED OR TRANSMITTED IN ANY FORM OR BY ANY MEANS, ELECTRONIC OR MECHANICAL, INCLUDING PHOTOCOPY, RECORDING, OR ANY INFORMATION STORAGE AND RETRIEVAL SYSTEM, WITHOUT PERMISSION IN WRITING FROM THE PUBLISHER.

ACADEMIC PRESS, INC.
111 Fifth Avenue, New York, New York 10003

United Kingdom Edition published by
ACADEMIC PRESS, INC. (LONDON) LTD.
24/28 Oval Road, London NW1

Library of Congress Cataloging in Publication Data

Main entry under title:

Optoacoustic spectroscopy and detection.

Includes bibliographies.
1. Optoacoustic spectroscopy. I. Pao, Yoh-han.
QD96.06067 535′.84 76-52728
ISBN 0-12-544150-9

PRINTED IN THE UNITED STATES OF AMERICA

Contents

List of Contributors

Numbers in parentheses indicate the pages on which the authors' contributions begin.

P. C. Claspy (133), Department of Electrical Engineering and Applied Physics, Case Western Reserve University, Cleveland, Ohio

C. Forbes Dewey, Jr. (47), Department of Mechanical Engineering, Massachusetts Institute of Technology, Cambridge, Massachusetts

Jerry A. Gelbwachs (79), Chemistry and Physics Laboratory, The Aerospace Corporation, El Segundo, California

P. L. Kelley (113), Lincoln Laboratory, Massachusetts Institute of Technology, Lexington, Massachusetts

Lloyd B. Kreuzer (1), Hewlett-Packard Laboratories, Palo Alto, California

M. B. Robin (167), Bell Laboratories, Murray Hill, New Jersey

*Allan Rosencwaig** (193), Bell Laboratories, Murray Hill, New Jersey

John D. Stettler (27), U.S. Army Missile Research, Development, and Engineering Laboratory, Redstone Arsenal, Alabama

Norman M. Witriol (27), U.S. Army Missile Research, Development, and Engineering Laboratory, Redstone Arsenal, Alabama

* Present address: Lawrence Livermore Laboratory, Livermore, California.

Preface

This book is an introduction to the principles and practice of optoacoustic spectroscopy and is intended to be of use in at least the following three contexts:

(1) as a treatise for self-study by researchers who wish to approach the subject on a broad intellectual basis;
(2) as a reference, containing many quantitative but simple results, considerable data, and an extensive bibliography; and finally
(3) as a guide, to serve as a basis for evaluation of the feasibility of using such techniques in specific instances.

The book consists of eight articles, organized in the form of eight independent chapters. These are original articles written by men who have contributed to the recent rapid development in this field. The articles vary in the extent to which they are expository or review, depending on the nature of the material covered.

In any first reading of this treatise, different groupings of chapters would be of greater interest and relevance, depending on the principal interest of the reader.

For example, in the early 1970s, the initial resurgence of interest in optoacoustic spectroscopy was primarily due to the high sensitivity attained in the use of the method for the detection of trace amounts of gas pollutants in the earth's atmosphere. The increased sensitivity was primarily due to the use of lasers as sources of excitation. The spectral region was that of infrared, and the physical mechanism responsible for the effect was that of vibrational–translational energy transfer in the gas. These types of results have been reproduced by many other investigators and the optoacoustic technique is by now probably the preferred method for detecting traces of gas contaminants at any level less than 1 part in 10^7 and for measuring gas adsorption coefficients with values less than 10^{-7} cm^{-1}.

In this book, the reader interested primarily in the detection and identification of gas contaminants would read Chapter 1 by Kreuzer to obtain a general introduction and a physical description of the generation and measurement of the optoacoustic signal. The description so obtained would be macroscopic and thermodynamic in nature. Chapter 2 by Stettler and Witriol provides an understanding of the optoacoustic effect on a molecular scale, with descriptions of the energy transfer processes and estimates of the lifetimes of vibrationally excited states. In Chapter 3 Dewey makes known the options available to the researcher in the choice of optoacoustic system design. He compares the capabilities and limitations of various optoacoustic system designs and presents quantitative relations for predicting system performance. Applications and specific results are discussed by Claspy in Chapter 6, with emphasis on the use of CO and CO_2 lasers in the 5–6 μm and 9–12 μm wavelength regions. With these two types of lasers, excitation may be achieved at a very large number of different discrete wavelengths. Although such lasers are tunable, they are not continuously tunable and cannot, for example, be used in the wavelength modulation mode; nor are they appropriate for molecules that absorb, say, in the 7 μm region. In the latter circumstance, in the infrared, either parametric oscillator or tunable diode lasers may be used and these are discussed by Kelley in Chapter 5.

On the other hand, the researcher interested in solid state optoacoustic spectroscopy might want to start off by reading Chapter 8 by Allan Rosencwaig. For the visible spectral region, good nonlaser light sources are available and Rosencwaig shows that very interesting results can be obtained even with incoherent light sources. In particular, the technique can be used for exploring the absorption spectra of substances such as powders, gels, adsorbed films, and even organic tissues. Such substances normally scatter so much light or are in a physical form such that absorption spectroscopy techniques cannot be used for investigative and characterization purposes. Optoacoustic spectroscopy promises to be very useful in this area and may prove to be even more so once tunable lasers are used for excitation. Laser sources and especially tunable lasers available for the ultraviolet and visible wavelength regions are discussed by Gelbwachs in Chapter 4. Photochemists and organic chemists are likely to find Chapter 7 by Robin on optoacoustically determined electronic spectra to be of particular interest.

To date, the optoacoustic effect has been employed in a wide variety of experimental situations, each having its particular needs with respect to sensitivity, accuracy, dynamic range, sample conditioning, and measurement. Accompanying these applications, there have been further significant advances in techniques, some examples of which are the use of second harmonic and wavelength modulation for discriminating against window noise and the use of resonant and/or multipath simple cells for increasing the signal level.

Heretofore, accounts of these advances and the many other results and discussions contained in this book could be found only in widely scattered technical journal literature. One of the purposes of this volume is to make these available to the researcher in a collected and coordinated format.

1

The Physics of Signal Generation and Detection

Lloyd B. Kreuzer

Hewlett–Packard Laboratories
Palo Alto, California

I. Introduction

In optoacoustic spectroscopy light energy is first converted into sound and then into an electrical signal (Hill and Powell, 1968). Figure 1a represents a simplified experimental arrangement. Light from the laser is modulated by passing it through a rotating "light chopper." The modulated beam then passes through a container that holds a gas sample. Energy absorbed by the gas from the beam heats the gas and causes its pressure to rise. Since the beam is modulated, this pressure rise is periodic at the beam

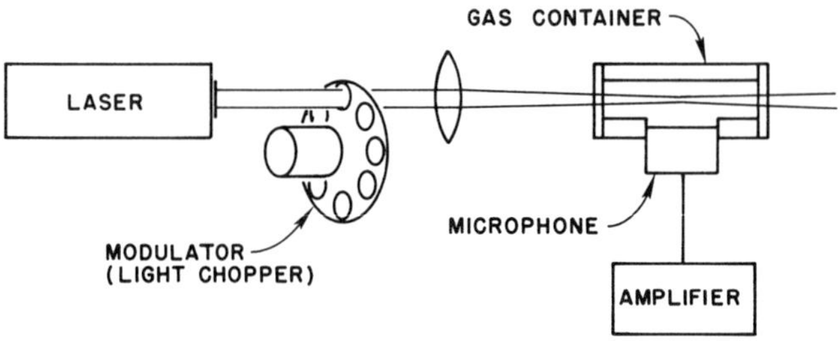

(a)

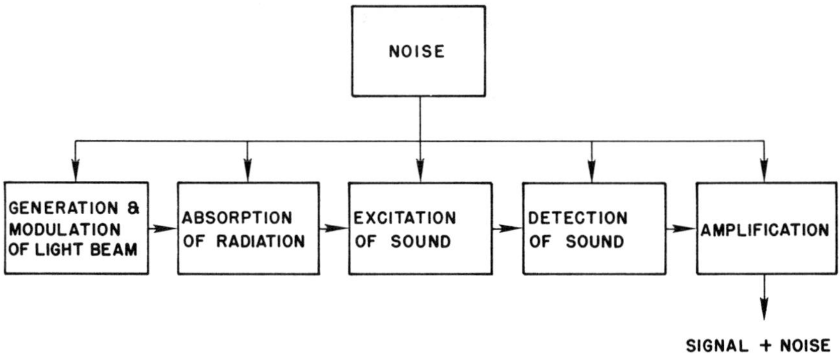

(b)

Fig. 1. Laser optoacoustic spectrometer. (a) Diagram of the apparatus. (b) The steps in the generation of the optoacoustic signal.

modulation frequency. It is detected by a microphone and converted into an electrical signal. The electrical signal is fed into the input of an amplifier. Figure 1b represents the stages of this process. At each stage of the process there will be mechanisms which add noise and degrade the ability to detect a small signal. The existence of noise sources is included in Fig. 1b.

This chapter provides a physical description of the generation and measurement of the optoacoustic signal. Noise sources that come from thermal fluctuations and noise at the input of the amplifier are discussed. The chapter concludes with a short section on optimum design and a numerical example.

II. Absorption of Light

The first step in the generation of the optoacoustic signal is the absorption of energy from the modulated light beam. This absorbed energy produces a periodically varying heat source in the gas that acts as the source of sound.

In this section the production of this heat source and its dependence on the properties of the incident beam and the absorbing gas is described.

Let the intensity in ergs $cm^{-2}\ sec^{-1}$ of the light beam be given by $I(\mathbf{r}, t)$, where $\mathbf{r}$ describes the position and t is the time, and let the heat source produced by this beam be $H(\mathbf{r}, t)$. H has dimensions of power per unit volume (erg $cm^{-3} \cdot sec$). In many common experimental situations, H and I are related by a simple proportionality constant α, which

$$H = \alpha I \tag{1}$$

has the dimensions of reciprocal length (cm^{-1}) and is called *absorbance*, as measured by the optoacoustic method. Equation (1) is valid when two conditions are satisfied simultaneously. First, the intensity I is sufficiently small so that the absorbing transition is not saturated. Second, the time variation of I is much slower than the rate of transfer of absorbed energy from the absorbing transition into heat. If the first condition is not satisfied, then the relationship between H and I becomes nonlinear, and H will contain frequency components not present in I. For example, if I has a sinusoidal time dependence of frequency ω, then H will have frequency components at ω and 2ω as well as other frequencies with amplitudes related in a nonlinear way to the amplitude of I. On the other hand, if the first condition is satisfied, but the second is not, then the time variation of H will lag behind that of I, and the constant α in Eq. (1) is replaced by a term that is frequency dependent. If the Fourier transforms of $I(\mathbf{r}, t)$ and $H(\mathbf{r}, t)$ are $I(\mathbf{r}, \omega)$ and $H(\mathbf{r}, \omega)$, then

$$H(\mathbf{r}, \omega) = \alpha(\omega) I(\mathbf{r}, \omega). \tag{2}$$

If both conditions are not satisfied, then the behavior can be very complex and quantum mechanical coherent effects may be important. Although this type of effect undoubtedly can be detected by optoacoustic means it has not been of importance, up to now. It is not discussed in this chapter. In the absence of quantum coherent effects, a rate equation treatment (Kaiser, 1959; Kreuzer, 1971) is adequate to connect I and H.

In order to illustrate the rate equation approach, a summary of equations describing a simple two-level system is given. Let N be the density of absorbing molecules, n_1 the density of absorbing molecules in the excited state, $h\nu$ the energy of the transition, $\Delta\nu$ the linewidth of the transition, S the line strength of the transition, τ_R the radiative lifetime, and τ_C the collisional decay time of the upper state. The equation describing upper state population follows in a straightforward manner assuming that the only paths for decay of the upper state are radiation and collision-induced decay:

$$\frac{dn_1}{dt} = -n_1\left[\frac{I}{h\nu}\left(\frac{S}{\pi\,\Delta\nu}\right) + \tau_C^{-1} + \tau_R^{-1}\right] + (N - n_1)\left(\frac{I}{h\nu}\right)\left(\frac{S}{\pi\,\Delta\nu}\right). \tag{3}$$

In Eq. (3), collisional excitation of the upper state has been disregarded. This is valid as long as $h\nu \gg kT$, where k is Boltzmann's constant and T is the gas temperature.

The solution of Eq. (3) gives the dependence of the upper state population on the light intensity I, the density of absorbing molecules N, and the various parameters that describe the transition. Let the time dependence of I be expressed in the form

$$I(t) = I_0[1 + f(t)] \tag{4}$$

where I_0 is a constant; then the solution of Eq. (3) is

$$n_1(t)/N = D/F + (2FT)^{-1}\left\{1 - F\exp\left\{-\int^t [F + 2Df(t')]\,dt'\right\}\right.$$
$$\left.\times \int^t \exp\left\{+\int^{t'} [F + 2Df(t'')]\,dt''\right\}dt'\right\}. \tag{5}$$

The spontaneous decay rate τ^{-1} of the upper state is given by

$$\tau^{-1} \equiv \tau_R^{-1} + \tau_C^{-1}. \tag{6}$$

The stimulated rate D for absorption or emission of radiation is given by

$$D \equiv I_0 S/(h\nu\pi\,\Delta\nu), \tag{7}$$

and the quantity F is defined by

$$F = 2D + \tau^{-1}. \tag{8}$$

The solution to Eq. (3) which gives the upper state population density can be simplified if the time variation of $I(t)$ is sufficiently slow. More precisely, if the frequency of variation of $I(t)$ is slow compared with F, then the left side of Eq. (3) may be considered equal to zero. The equation then may be solved for n_1:

$$\frac{n_1}{N} = \frac{(IS/h\nu\pi\,\Delta\nu)}{(2IS/h\nu\pi\,\Delta\nu) + \tau^{-1}}. \tag{9}$$

This shows the transition saturation for large I ($2IS/h\nu\pi\,\Delta\nu \gg \tau^{-1}$). In the limit of large I, n_1/N becomes equal to $\frac{1}{2}$, and the absorbing molecules are evenly divided between the upper and lower states of the transition. In the limit of small I, Eq. (9) reduces to

$$\frac{n_1}{N} \sim \frac{IS\tau}{h\nu\pi\,\Delta\nu}. \tag{10}$$

Thus, when the time variation of I is sufficiently slow and the intensity sufficiently weak, then n_1 is proportional to I.

Heat is generated in the gas by the nonradiative decay of the excited state population n_1. The rate of heat generation is given by

$$H = (n_1/\tau_C)h\nu. \tag{11}$$

In those cases when Eq. (10) is a good approximation, then Eqs. (10) and (11) may be combined to give the dependence of H on I. Doing this and comparing the result to Eq. (1) leads to

$$\alpha = N\frac{S\tau}{\pi\,\Delta\nu\tau_C}, \tag{12}$$

which gives α as a function of the constants that describe the transition. The optoacoustic absorbance given by Eq. (12) describes the conversion of light energy into heat. Some of the absorbed energy may, of course, follow other paths. For example, it may be reradiated or may cause photochemical reactions. Convention absorption spectroscopy, which measures the attenuation of light as it passes through the sample, measures the extinction coefficient which represents the effect produced by all processes which remove energy from the light beam. Comparison of conventional and optoacoustic absorption measurements can give information on the type of absorption processes present. For example, it has been shown (Harshberger and Robin, 1973) that the two measured absorptions can differ by the amount of absorbed energy that goes into driving a chemical reaction.

III. Excitation of Sound

The second step in the generation of the optoacoustic signal is the excitation of sound in the gas. The details of this process are described in this section (Kerr and Atwood, 1968; Goldan and Gato, 1974; Dewey *et al.*, 1973; Dewey, 1974; Rosengren, 1975).

First, the excitation of the acoustic normal modes of the gas container is described. Then the calculation of the Q of these modes is outlined and finally the generation of noise by thermal fluctuations is discussed.

A. *Normal Mode Amplitudes*

Sound in the gas can be described by the acoustic pressure $p(\mathbf{r}, t)$ which is the difference between the total pressure P and its average value P_0:

$$p = P - P_0. \tag{13}$$

Associated (Morse and Ingard, 1961) with acoustic pressure p is an acoustic velocity $\mathbf{u}(\mathbf{r}, t)$ and an acoustic temperature $\Theta(\mathbf{r}, t)$. The acoustic velocity is the fluid velocity of the gas at position $\mathbf{r}$ and time t caused by the sound. The acoustic temperature is the departure from the average temperature T caused by the sound.

The heat $H(\mathbf{r}, t)$ produced by the absorption of light acts as a source for the generation of sound (Morse and Ingard, 1961). This is described by

$$\nabla^2 p - c^{-2}\,\partial^2 p/\partial t^2 = -[(\gamma - 1)/c^2]\,\partial H/\partial t, \tag{14}$$

where c is the velocity of sound and γ is the ratio of the specific heat of the gas at constant pressure C_P to that at constant volume C_V. This equation does not include the effects of acoustic loss produced by heat conduction and viscosity. In order to properly discuss the optoacoustic effect, it is necessary to include these mechanisms. Although this could be done by modifying Eq. (14), the resulting equation and its solution are unnecessarily complex. The approach followed here consists of including the loss as a perturbation of the loss-free solutions of Eq. (14).

Equation (14) is an inhomogeneous wave equation that may be solved in the usual manner by taking the time Fourier transform of both sides and expressing the solution p as an infinite series expansion of the normal mode solutions p_j of the homogeneous wave equation. The Fourier transform of Eq. (14) is given by

$$(\nabla^2 + \omega^2/c^2)p(\mathbf{r}, \omega) = [(\gamma - 1)/c^2]i\omega H(\mathbf{r}, \omega), \tag{15}$$

where

$$p(\mathbf{r}, t) = \int p(\mathbf{r}, \omega)e^{-i\omega t}\,d\omega, \tag{16a}$$

$$H(\mathbf{r}, t) = \int H(\mathbf{r}, \omega)e^{-i\omega t}\,d\omega. \tag{16b}$$

The normal mode solutions of the homogeneous wave equation are determined by the boundary conditions. If the walls of the gas container are rigid, the acoustic velocity component normal to the wall must vanish at the wall. Since the acoustic velocity $\mathbf{u}$ is proportional to the gradient of p as expressed by

$$\mathbf{u}(\mathbf{r}, \omega) = (i\omega\rho_0)^{-1}\,\nabla \cdot p(\mathbf{r}, \omega) \tag{17}$$

it follows that the gradient of p normal to the boundary must vanish at the boundary. This boundary condition determines the normal mode solutions p_j of the homogeneous wave equation

$$(\nabla^2 + k_j^{\,2})p_j(r) = 0. \tag{18}$$

Let the resonant frequency of the normal mode $p_j(r)$ be ω_j. These modes will be orthogonal and may be normalized with the normalization condition given by

$$\int p_l{}^* p_j \, dV = V_c \, \delta_{lj}. \tag{19}$$

The volume integral is over the volume V_c of the gas container.

If, for example, the gas container is a cylinder of radius a, and length l, then it is convenient to write Eq. (18) in cylindrical coordinates (Morse, 1948):

$$r^{-1}\frac{\partial}{\partial r}\left(r\frac{\partial p_j}{\partial r}\right) + r^{-2}\frac{\partial^2 p_j}{\partial \phi^2} + \frac{\partial^2 p_j}{\partial z^2} + \frac{\omega_j{}^2}{c^2} p_j = 0. \tag{20}$$

The solution (Morse, 1948) of this equation is given by

$$p_j = \begin{matrix}\cos\\ \sin\end{matrix}(m\phi)[AJ_m(k_r r) + BN_m(k_r r)][C\sin(k_z z) + D\cos(k_z z)], \tag{21}$$

where J_m and N_m are Bessel functions of the first and second kind, respectively. Since N_m becomes infinite as r approaches zero, it follows that $B = 0$, if Eq. (20) is to represent the pressure inside a cylinder shaped container. In order to satisfy the boundary condition imposed by the rigid gas container, the gradient of p normal to the container walls must vanish at the walls. If one end of the gas container is at $z = 0$, and the other end is at $z = l$, then it follows that $C = 0$ and the allowed values of k_z are given by

$$k_z = (\pi/l)n_z \qquad (n_z = 1, 2, 3, \ldots). \tag{22}$$

Applying the same boundary condition to the walls at $r = a$ leads to the condition that the derivative of $J_m(k_r r)$ with respect to r must vanish at $r = a$:

$$[dJ_m(k_r r)/dr]\Big|_{r=a} = 0. \tag{23}$$

This is equivalent to

$$[dJ_m(\pi\alpha)/d\alpha]\Big|_{\alpha=\alpha_{mn}} = 0, \tag{24a}$$

$$k_r = \pi\alpha_{mn}/a, \tag{24b}$$

where α_{mn}, the nth root of the equation involving the mth order Bessel function, has been tabulated (Morse, 1948). The requirement that p be continuous limits m to integral values. Substitution of Eq. (21) into Eq. (20) gives the resonant frequency of the mode:

$$\omega_j = c\sqrt{k_z{}^2 + k_r{}^2}. \tag{24c}$$

If the gas container has, for example, a length $l = 7$ cm, a radius $a = 0.5$ cm, and the speed of sound is $c = 3.31 \times 10^4$ cm sec^{-1}, then the normal modes and resonant frequencies can be calculated from the above equation and tabulated values of α_{mn}. The lowest-order longitudinal mode ($n_z = 1$, $m = 0$, $n = 0$) has a resonant frequency of 2.37 kHz, while the lowest-frequency radial mode ($n_z = 0$, $m = 1$, $n = 0$) has a resonant frequency of 19.4 kHz.

The acoustic pressure p can be expressed as an expansion over the modes p_j with mode amplitudes A_j:

$$p(\mathbf{r}, \omega) = \sum_j A_j(\omega) p_j(\mathbf{r}). \tag{25}$$

Substituting Eq. (25) into Eq. (14) and making use of Eqs. (18)–(20), gives

$$A_j(\omega) = -\frac{i\omega}{\omega_j^2} \frac{[(\gamma - 1)/V_c] \int p_j^* H \, dV}{(1 - \omega^2/\omega_j^2)}, \tag{26}$$

which gives the mode amplitudes. The integral in the numerator on the right side of Eq. (26) represents the coupling between the heat source H and the normal mode p_j. The denominator represents the mode resonance with A_j becoming infinite as ω approaches the natural resonant frequency ω_j. This physically unreasonable situation is the result of the absence of any loss mechanism in Eq. (14). It may be corrected by modifying Eq. (26) to include a mode damping described by the quality factor Q_j:

$$A_j(\omega) = -\frac{i\omega}{\omega_j^2} \frac{[(\gamma - 1)/V_c] \int p_j^* H \, dV}{[1 - \omega^2/\omega_j^2 - i\omega/\omega_j Q_j]}. \tag{27}$$

The method of calculating Q_j from the undamped function p_j is discussed below.

In order to show explicitly the dependence of the acoustic signal on the gas absorption and the light intensity, it is necessary to replace H in Eq. (27). Under those conditions in which Eq. (1) is a good approximation, this substitution is very easy and Eq. (27) becomes

$$A_j(\omega) = -\frac{i\omega}{\omega_j^2} \frac{\alpha[(\gamma - 1)/V_c] \int p_j^* I \, dV}{[1 - \omega^2/\omega_j^2 - i\omega/\omega_j Q_j]}. \tag{28}$$

In order to understand Eq. (28), it is useful to consider two special cases. First, assume that I is constant throughout the volume of the container. In this case, the integral in the numerator of Eq. (26) vanishes for $j \neq 0$. The lowest-order mode p_0 has a resonant frequency of $\omega_0 = 0$ and represents a constant pressure change in the container which is independent of position. In this case I and p_0 are proportional to each other and the vanishing of the integral for $j \neq 0$ is a direct result of the orthogonality of the functions p_j. Under these conditions, Eq. (28) reduces to

$$A_0(\omega) = \frac{i\alpha(\gamma - 1)I}{\omega[1 + (i/\omega\tau_T)]}. \tag{29}$$

The time τ_T in the denominator of Eq. (29) is the damping time of p_0 resulting from heat conduction from the gas to the container walls. Assuming that the gas container is cylindrical in shape with cross section area A_c, length l, and volume V_c, then the total light beam power W is equal to IV_c/l. The intensity I in Eq. (29) can therefore be replaced by Wl/V_c, which gives

$$A_0(\omega) = \frac{i\alpha(\gamma - 1)Wl}{\omega(1 + i/\omega\tau_0)V_c}. \tag{30}$$

Equation (30) was derived from the assumption that the light was uniform and filled the entire gas container. It is however also valid when the intensity does not fill the entire container, although in this case, other mode amplitudes will also be excited.

As a second special case, consider the situation where spatial distribution of the intensity is adjusted so that only the first-order mode p_1 is excited. Then Equation (28) reduces to

$$A_1(\omega) = -\frac{i\omega}{\omega_1{}^2} \frac{\alpha(\gamma - 1)Wl}{V_c[1 - (\omega^2/\omega_1{}^2) - i(\omega/\omega_1 Q_1)]}. \tag{31}$$

In general, it may not be possible to adjust I so that only the first mode is excited, and Eq. (31) should be regarded as the upper limit on mode amplitude.

To understand the advantages of exciting a resonant mode ($i \neq 0$) or the zero order mode, it is useful to compare the amplitudes given by Eqs. (30) and (31). At low frequency, when $\omega\tau_T \ll 1$, the zero-order mode amplitude is constant, independent of frequency, and has the value $A_0 = \alpha(\gamma - 1)Wl\tau_T/V_c$. At high frequencies $\omega\tau_T \gg 1$, it decreases as ω^{-1}. The first-order mode amplitude reaches a maximum at $\omega = \omega_1$. The ratio of the maximum amplitude of the first resonant mode to that of the zero order mode is given by

$$A_1(\omega_1)/A_0(0) = Q_1/\omega_1\tau_1. \tag{32}$$

If the value of this ratio is greater than unity, then the first-order mode amplitude can be made larger than the zero-mode. The value of this ratio depends on mode damping, which is discussed in the next section.

B. Calculation of Q

In the previous discussion, the physically unreasonable infinite Q present in Eq. (26) was corrected by assigning a finite Q to each mode. This finite Q is usually the result of loss caused by heat conduction and viscosity. However, under some circumstances, other losses may be important. For example, the finite time required for rotational and vibrational degrees of freedom of the gas molecules to come to equilibrium with translational degrees of freedom can also introduce loss. Only loss from heat conduction and viscosity is discussed further in this chapter.

Loss produced by viscosity and heat conduction can be separated into a volume loss and a surface loss (Morse and Ingard, 1961). The surface loss occurs in a thin region near the walls. This region may be considered to be composed of two layers that extend out from the wall. One of thickness l_v in which the viscosity effects take place and one of thickness l_h in which heat conduction effects occur. These skin depths are given by

$$l_v^2 = 2\eta/\omega\rho_0, \tag{33}$$

$$l_h^2 = 2\kappa/\rho_0\omega C_P. \tag{34}$$

In these equations, η is the viscosity and κ the thermal conductivity of the gas. The nature of these layers can be understood in the following manner. The container wall will have thermal conductivity much greater than the gas and thus near the wall the gas will have a constant temperature and expand and contract in an isothermal manner. Far from the walls, the expansion and contraction caused by the sound is very nearly adiabatic. Acoustic loss from heat conduction occurs in the region where the gas expansion is partly adiabatic and partly isothermal. At the wall surface, the component of the acoustic velocity u_t tangent to the wall surface vanishes owing to viscosity. Far from the wall, the acoustic velocity is proportional to the gradient of p and is given by Eq. (17). Loss occurs owing to viscosity in the region near the walls where the transition from the condition of Eq. (17) to the condition $u_t = 0$ occurs. The total surface loss L_s is given by

$$L_{s_j} = |A_j|^2 \int [\tfrac{1}{2}R_v|u_{tj}|^2 + \tfrac{1}{2}R_h|p_j|^2]\, dS, \tag{35}$$

where the integration is over the surface of the container and R_v and R_h, which represent the loss from viscosity and heat conduction are given by

$$R_v \equiv \sqrt{\eta\omega\rho_0/2}, \tag{36}$$

$$R_h = [(\gamma - 1)/\rho_0 c^2]\sqrt{\kappa\omega/2\rho C_V}. \tag{37}$$

In Equation (35) u_{tj} is the component of acoustic velocity tangential to the walls that would be present if there were no viscosity or heat conduction.

The volume loss L_v is given by

$$L_{v_j} = (\omega_i^2/\rho^2 c^4)[(\gamma - 1)(\kappa/2C_P) + (2n/3)]V_c|A_j|^2 \tag{38}$$

The energy E_j stored in mode j is given by

$$E_j = V_c|A_j|^2/\rho_0 c^2. \tag{39}$$

The Q of the mode can easily be calculated from Eqs. (35), (38), (39), and the definition of Q:

$$Q_j = \omega_j \frac{\text{energy stored in mode } j}{\text{rate of loss of energy from mode } j}, \tag{40}$$

$$Q_j = \omega_j \frac{E_j}{L_{s_j} + L_{v_j}}. \tag{41}$$

C. *Thermal Fluctuation Noise*

The acoustic normal modes of the gas-filled cavity are also excited by thermal fluctuations. This produces a noise source that causes a fundamental limitation on acoustic signal detection sensitivity. This situation is described by the theory of thermal fluctuations (Kittel, 1958). Each mode has an average energy of kT. The power spectrum of the noise is given by

$$|A_{jn}(\omega)|^2 = \frac{4\rho_0 c^2 kT}{V_c \omega_j Q_j[(1 - \omega^2/\omega_j^2)^2 + (\omega/\omega_j Q_j)^2]}. \tag{42}$$

Mode damping plays an important and often misunderstood role in thermal fluctuation noise. The total energy of excitation of a mode is equal to kT and is not affected by Q_j. However, the frequency composition of the noise depends on Q_j. Although the shape of $|A_{jn}(\omega)|^2$ changes with Q_j, the integral of this function over frequency, which gives the mean energy of excitation, remains constant. Since the noise energy in a frequency band of width $\Delta\omega$ centered at ω is given by $|A_{jn}(\omega)|^2 \Delta\omega/\rho c^2$ it is evident that as Q_j is increased, the noise energy moves into a band around the resonant frequency ω_j. Increasing Q_j shifts the noise into a narrow band around the mode resonant frequency. In the case in which the signal frequency is well below the mode resonant frequency ($\omega \ll \omega_j$), Eq. (42) reduces to

$$|A_{jn}(\omega)|^2 = 4\rho_0 c^2 kT/\omega_j Q_j V_c. \tag{43}$$

This shows that in the case in which the signal frequency is well below the mode resonant frequency that the noise may be reduced by increasing Q_j.

It is often useful to evaluate the noise equivalent power (NEP) of a detector as a means of evaluating its performance. The NEP of an optoacoustic detector is the amount of power that would have to be absorbed by the gas to produce a signal amplitude equal to the noise amplitude given by Eq. (42). It can be calculated from Eqs. (28) and (42). The result of this calculation for a frequently encountered situation is presented next.

It is often the case that the signal is generated in the zero-order mode. As described previously this occurs when the light intensity is constant throughout the gas container. Under this condition, Eq. (30) gives the signal amplitude. In general, an optoacoustic system is operated at high enough frequency that $\omega\tau_T \gg 1$, and under this condition Eq. (30) simplifies to

$$A_0(\omega) = \frac{i\alpha(\gamma - 1)Wl}{\omega V_c}. \tag{44}$$

The absorbed power that generates the signal is αWl. The fluctuation noise of the zero-order mode is zero as can be seen from Eq. (42) by taking the limit as ω_j goes to zero. However, noise from other modes will interfere with any attempt to measure the mode amplitude. The noise may come predominately from the first-order mode in which case the NEP can be found by equating the absolute value of the right side of Eq. (44) with the right side of Eq. (43), and replacing αWl by NEP. Solving this equation for NEP gives

$$(\mathrm{NEP})^2 = 8\pi\rho_0 c^2 kTV_c\omega^2/\omega_1 Q_1(\gamma - 1)^2. \tag{45}$$

A factor of 2π has been included on the right side of Eq. (45) in order to give NEP in the usual engineering units of power per root hertz.

Equation (45) can be used to optimize the design of an optoacoustic system. It shows that the NEP is reduced by increasing the resonant frequency and Q. It also provides a means of evaluating the performance of real systems that will have other noise sources in addition to the thermal fluctuation noise described by Eq. (45). Thus, any real system will have a noise equivalent power greater than or equal to the value given by Eq. (45).

IV. Signal Detection

The third and final step in the generation of the optoacoustic signal is the detection of the acoustic excitation. The detector is usually a condenser microphone, and this situation is treated in this section. In order to be able to calculate the size of the electrical signal produced by the microphone and understand the noise that will be added to the signal by the microphone and amplifier, it is necessary to understand the behavior of the microphone. Two different models of microphone behavior are useful. Both give the same results but since they take different viewpoints, each is useful for a complete understanding. Both models are presented in summary. The first, or mechanical, model treats the microphone mechanically in terms of the position, mass, damping, and spring constant of the diaphragm. The second, or electrical, model (Becking and Rodenakers, 1954; Morse and Ingard, 1948) treats the microphone electrically in terms of an equivalent circuit. The discussion will be restricted to a condenser microphone.

A. Microphone Mechanical Model

The diaphragm of a condenser microphone is usually made of thin metal that is mounted in such a manner that a large radial tension is produced. Acoustic pressure acting on one side of the diaphragm causes it to move. This motion is detected by a change in capacity between the diaphragm and a fixed plate, which is mounted behind the diaphragm. Motion of the diaphragm may be described by the modes of vibration of a thin plate. Since only the lowest-order mode will cause a significant change in capacity, it is only necessary to consider this mode. The motion of the diaphragm that generates a signal can thus be described by a single degree of freedom. This corresponds to the diaphragm bending into a spherical shape. Assume that when the diaphragm is at rest, it is in the yz plane with its center at the origin. Then, the lowest-order mode corresponds to a displacement of each point of the diaphragm in the x direction by an amount $x(r)$ given by

$$x(r) = x(0)[1 - r^2/a^2], \qquad (46)$$

where r is the distance of the diaphragm point from the origin, $x(0)$ is the displacement of the diaphragm at the origin, and a is the radius of the diaphragm. The average displacement of the diaphragm, designated here by x is found by averaging $x(r)$ over the diaphragm:

$$x = (\pi a^2)^{-1} \int\int x(r) r \, dr \, d\Theta. \qquad (47)$$

Substituting Eq. (46) into Eq. (47) gives

$$x = \tfrac{1}{2} x(0). \qquad (48)$$

The average motion is one-half the motion at the center. The equation of motion in the average coordinate x for the lowest-order mode is

$$m \, d^2x/dt^2 + \delta \, dx/dt + K_m x = F, \qquad (49)$$

where m is the mass of the diaphragm, δ the damping, K_m the restoring force, and F the external force applied to the diaphragm. The external force is composed of two parts: (1) the force resulting from the sound pressure $p_m A_m$, where p_m is the average of the pressure over the diaphragm and A_m is the area of the diaphragm, and (2) the force resulting from the bias voltage.

The force caused by the bias voltage causes the equilibrium position of the diaphragm to shift by a small amount x_0, which is given by

$$x_0/d = C_m V_B^2/d^2 K_m, \tag{50}$$

where d is the distance between the diaphragm and the back plate of the microphone capacitor when the microphone is unbiased, C_m is the unbiased microphone capacity given by

$$C_m = \varepsilon_0 A_m/d, \tag{51}$$

and V_B is the microphone bias voltage.

The restoring force K_m is produced by the tension T_m in the microphone diaphragm:

$$K_m = 8\pi T_m. \tag{52}$$

This tension must be sufficient to prevent the bias voltage from pulling the diaphragm into contact with the back plate. It is useful to define a microphone equivalent volume V_m, which is a function of T_m:

$$V_m = \frac{\gamma P_0 A_m^2}{8\pi T_m}. \tag{53}$$

If the diaphragm was not tensioned, but instead was stretched like a drum skin over a volume V_m of air, then the compression of this air would produce the same restoring force as the tension T_m. The microphone thus behaves as if the diaphragm were not tensioned and instead a volume V_m of air in a closed container behind the diaphragm provided the restoring force. It is shown below that this model is useful in understanding the coupling of the microphone to the gas container.

There are various methods of biasing a condenser microphone. Each is useful under different circumstances. The charge on the microphone may remain constant as the diaphragm moves and the signal be detected as a voltage by a high-impedance amplifier. Alternatively, the voltage may remain constant and the signal be detected as a current by a low impedance current-sensitive amplifier. Alternating current bias schemes have been utilized in which the motion of the diaphragm and the resultant change in capacity either frequency modulates a carrier, or unbalances a bridge circuit. Only the first bias method is discussed in this chapter. This scheme of constant charge bias is most easily achieved by connecting the microphone to a bias voltage through a large resistor R_B, which is sufficiently large so that $(R_B C_m)^{-1}$ is much less than the signal frequencies of interest. It is easily shown (Olson, 1947) that the output voltage, V_s, produced by the microphone at the input of a high-impedance amplifier under the above conditions is given by

$$V_s = p_m \frac{V_B A_m}{d K_m} \left[1 - \frac{\omega^2}{\omega_m{}^2} - i \frac{\omega}{\omega_m Q_m} \right]^{-1}, \tag{54}$$

where ω_m is the microphone resonant frequency $\omega_m \equiv \sqrt{K_m/m}$ and Q_m the quality factor given by $Q_m = \sqrt{mK_m/\delta}$. The open-circuit voltage sensitivity S_m is defined as the low-frequency ($\omega \ll \omega_m$) ratio of signal voltage v_s to sound pressure p_m:

$$S_m = V_B A_m / d K_m. \tag{55}$$

Combining Eqs. (55), (53), and (52) gives S_m as a function of the microphone equivalent volume V_m:

$$S_m = V_B V_m / d \gamma P_0 A_m. \tag{56}$$

The electrical signal generated by the microphone as a result of the optoacoustic pressure signal can be calculated from Eq. (54) by substituting the proper value for p_m. In solving Eq. (14) to find p_m it is necessary to consider the effect of the microphone on the acoustic behavior of the gas inside its container. The addition of the microphone to the container affects the acoustic modes of the container by changing the boundary conditions on p_j. In the absence of the microphone, the boundary condition produced by the rigid container walls requires that the gradient of p normal to the wall must vanish at the walls. When the microphone is added to the container, the diaphragm forms a part of the container wall and this part is no longer rigid. The boundary conditions require the acoustic velocity normal to the surface to be equal to the diaphragm velocity dx/dt produced by the acoustic pressure p_m. This gives

$$\left[\frac{(\nabla p)_N}{p} \right]_m = \frac{A_m}{K_m} \frac{\omega^2 \rho_0}{[1 - (\omega/\omega_m)^2 - i(\omega/\omega_m Q_m)]}, \tag{57}$$

which is the boundary condition at the microphone. In Eq. (57) $(\nabla p)_N$ is the gradient of p normal to the diaphragm. When the microphone is added to the gas container and Eq. (57) describes the boundary condition on p, then the solutions of Eq. (18) will not in general be orthogonal, and Eqs. (20), (25), and (26) will not be valid. Solving Eq. (18) for the optoacoustic pressure becomes more complex. However, the case in which the microphone is coupled to a small gas container can be treated in a simplified manner.

If the container is small enough and the modulation frequency is much less than both the container's first resonant frequency and the microphone resonance, then pressure will be constant independent of position in the container and the microphone will behave like an additional volume V_m added to the container. The pressure in the container is given by Eq. (30) with V_c replaced by $(V_c + V_m)$. The addition of the microphone reduces the pressure by an amount $V_c/(V_c + V_m)$ from what it would be without the microphone:

$$A_0(\omega) = \frac{i\alpha(\gamma - 1)Wl}{\omega(1 + i/\omega\tau_T)(V_c + V_m)}. \tag{58}$$

Comparing Eq. (58) with Eqs. (54) and (56) and assuming that $\tau_T^{-1} \ll \omega \ll \omega'$ gives

$$V_s = \frac{i(\gamma - 1)}{\omega\gamma P_0 A_m}\frac{V_B}{d}\left[\frac{V_m}{V_c + V_m}\right](\alpha W l), \tag{59}$$

which gives the electrical signal produced by the microphone. This equation describes the conditions encountered in many optoacoustic detectors. It provides a means of evaluating the effect of design changes on signal amplitude. It will be used in the section on optimizing design as part of the basis for obtaining optimum signal-to-noise performance.

If the dimensions of the gas chamber are sufficiently small, then the first acoustic resonance of the combined microphone and container will occur at the modified microphone resonance frequency. Coupling the microphone to the container modifies its resonant frequency by increasing the effective spring constant which acts on the diaphragm. When the diaphragm moves, it compresses air in the gas container which causes a restoring force. The modified spring constant K_m' can be calculated from the modified effective volume V_m':

$$(V_m')^{-1} = V_m^{-1} + V_c^{-1}, \tag{60}$$

$$K_m' = K_m[1 + V_m/V_c]. \tag{61}$$

The resonant frequency then becomes

$$\omega_1 = \omega_m\sqrt{1 + V_m/V_c}. \tag{62}$$

Assuming that the loss comes only from the microphone damping, then the noise voltage $|V_{sn}(\omega)|^2$ is calculated by combining Eq. (43), (54), (60), (61), and (62) noting that $\rho_0 c^2 = \gamma P_0$:

$$|V_{sn}(\omega)|^2 = \frac{4KT\rho_0 c^2}{\omega_m Q_m V_m(1 + V_m/V_c)^2} S_m^2. \tag{63}$$

This equation is also used in the consideration of optimum design.

Another easily treated case is the weak coupling of the microphone and the gas container. If the microphone is weakly coupled to the gas container, then it will not affect the normal modes of the container. In this situation, Eqs. (20), (25), (26), and (27) will be valid, and the normal mode expansion of p, averaged over the microphone diaphragm, may be substituted into Eq. (54):

$$V_s = -\frac{i\omega\alpha(\gamma - 1)}{V_c}\left\{\sum_j \frac{\int p_j^* I \, dV}{\omega_j^2[1 - \omega^2/\omega_j^2 - i(\omega/\omega_j Q_j)]}\right\} \times \frac{p_{im} S_m}{[1 - \omega^2/\omega_m^2 - i(\omega/\omega_m Q_m)]}. \quad (64)$$

The noise may be calculated in a similar manner from Eq. (42).

B. *Microphone Electrical Model*

A microphone may be viewed as a two-terminal electrical network. The small signal electrical impedance of a biased microphone is identical to a series resistance–inductance–capacitance (RLC) resonant circuit shunted by a capacitor. Viewed as a black box with two terminals, the microphone is indistinguishable from the equivalent circuit shown in Fig. 2. The signal

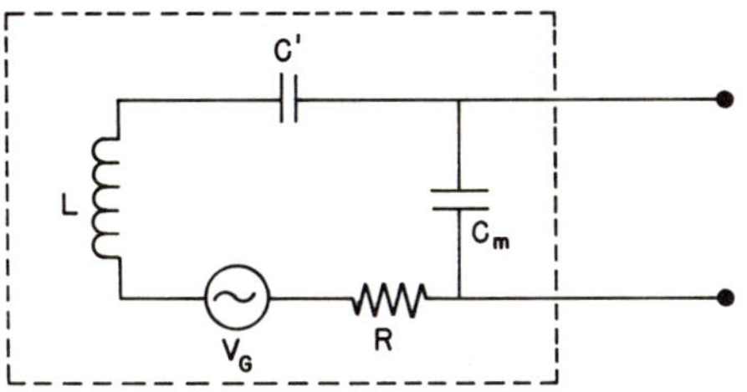

Fig. 2. Condenser microphone equivalent circuit. The above circuit is electrically equivalent to a biased condenser microphone, where C_m is the microphone high-frequency capacitance, C', L, and R are determined respectively by the restoring force, mass, and damping of the diaphragm:

$$C' = C_m^2 V_B^2/d^2 K_m, \qquad R = \delta/K_m C' = (Q_m w_m C')^{-1},$$
$$L = m/K_m C' = (\omega_m^2 C')^{-1}, \qquad V_G = (A_m d/V_B C_m)p = (C_m/C')S_m P_m.$$

The signal caused by an acoustic pressure p_m on the diaphragm appears as a voltage source V_G in series with the RLC' resonant circuit. The Johnson noise voltage generated in the resistence R appears on the output terminals as a thermal fluctuation noise caused by the microphone damping.

generated by acoustic pressure p_m on the diaphragm appears to come from series voltage source of magnitude V_G:

$$V_G = \frac{A_m d}{V_B C_m} p_m, \tag{65}$$

where C_m is the high-frequency microphone capacity given by

$$C_m = \varepsilon_0 A_m/d. \tag{66}$$

The resonant frequency and Q of the series circuit are equal to the corresponding values for the mechanical motion of the diaphragm. The capacitor C' has the value

$$C' = (x_0/d)C_m \tag{67}$$

which gives

$$C' = C_m{}^2 V_B{}^2/d^2 K_m \tag{68}$$

when Eq. (50) is used to replace x_0/d. In order to obtain the proper Q and resonant frequency, it follows that

$$L = (\omega_m{}^2 C')^{-1} \tag{69}$$

and

$$R = (Q_m C' \omega_m)^{-1}. \tag{70}$$

It is easily seen that the above values result in the generator voltage given by Eq. (65), producing the same output as given in Eq. (54). The electrical circuit model not only describes the signal properties of the microphone, but it also describes its noise properties.

In the previous discussion of noise, the noise appeared to be caused by viscosity and heat conduction which caused the finite Q of the mode. In the electrical model, the noise is represented by the resistor Johnson noise. In both models, it is the loss that is responsible for the noise. Johnson noise produced by the resistor in the equivalent circuit appears as a noise voltage on the output terminals that is exactly the same amplitude as the thermal fluctuation noise discussed above.

In calculating the microphone electrical noise, it is necessary to consider the effect of coupling of the microphone to the gas container. This coupling increases the effective spring constant of the diaphragm which, in the electrical model, results in a reduction of C' by an amount $(1 + V_m/V_c)^{-1}$. The noise at the output terminals of the microphone generated by the Johnson noise in resistor R is given by

$$|V_n|^2 = \frac{4kTR}{\omega^2 C_m{}^2}\left\{R^2 + \left(\omega L - \frac{[1 + V_m/V_c]}{\omega C'}\right)^2\right\}^{-1}. \tag{71}$$

In the low-frequency limit, this equation reduces to

$$|V_n|^2 = \frac{4kT\rho c^2}{\omega_m Q_m V_m(1 + V_m/V_c)^2} S_m{}^2, \tag{72}$$

which is identical to Eq. (63). Thus, both the electrical and mechanical model give the same value for the noise.

C. *Electrical Noise*

In addition to the thermal fluctuation noise discussed above, there may also be significant noise sources in the amplifier (Motchenbacher and Fitchen, 1973) connected to the microphone. Representing the microphone by the equivalent circuit of Fig. 2 is useful in discussing amplifier noise. A model of the microphone and amplifier that includes noise sources is given in Fig. 3. Amplifier noise is represented by a series voltage noise source $|V_{na}|^2$, and a shunt current noise $|I_{na}|^2$ source. Noise is also generated by Johnson noise of the bias resistor R_B. Capacitor C_2 and the associated feedback circuit is used to increase input impedance and thus increase low-frequency response. Combining these three noise sources with the microphone noise and assuming that $R_B \gg (\omega C)^{-1}$ gives the power spectrum

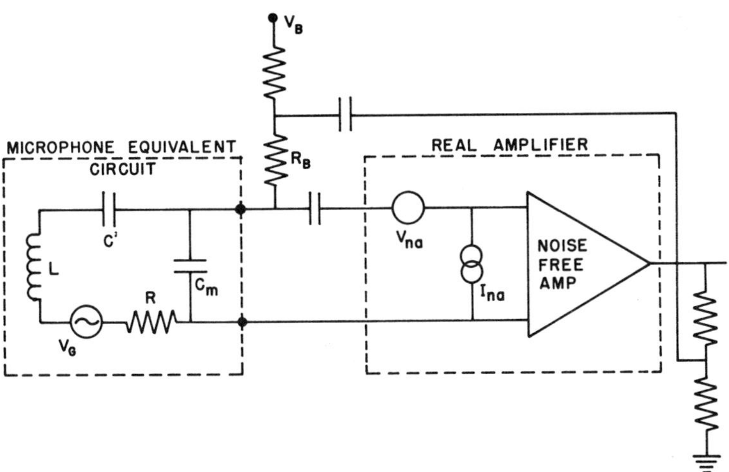

Fig. 3. Noise model of microphone and amplifier. The microphone is represented by the equivalent circuit of Fig. 2 and the amplifier is represented by an ideal noise-free amplifier with a series voltage noise source and a shunt current noise source. Thermal fluctuation noise comes from Johnson noise in the bias resistor R_B and the microphone damping resistance R. When the microphone is coupled to the gas container, the model must be modified by replacing C' by $C'/(1 + V_m/V_c)$. This change represents the increased stiffness of the diaphragm caused by the gas container volume V_c.

$|V_N|^2$ of the total noise referred to the amplifier input,

$$|V_N|^2 \approx |V_{na}|^2 + \frac{|I_{na}|^2 + 4KT/R_B}{\omega^2 C_m{}^2}$$

$$+ \frac{4KTC'}{Q_m \omega_m C_m{}^2 (1 + V_m/V_c)^2} [(\omega^2/\omega_1{}^2 - 1)^2 + (\omega/\omega_1 Q_1)^2]^{-1}. \quad (73)$$

The four sources of noise included in Eq. (73) produce three terms, each with a different frequency dependence. Amplifier voltage noise appears as a frequency-independent term. (Of course, if $|V_{na}|^2$ depends on frequency, this must be considered, but in general if ω is above the f^{-1} noise region, $|V_{na}|^2$ will be independent of frequency.) The second term that decreases with increasing frequency represents noise resulting from amplifier current noise and bias resistor noise. The bias resistor noise appears as an additional current noise of amplitude $4kT/R_B$. The third term, which is the microphone thermal noise, is constant at low frequency ($\omega \ll \omega_1$) goes through a resonance at ω_1, and decreases at high frequencies ($\omega \gg \omega_1$). At low frequencies the dominant noise source will be amplifier current noise and bias resistor noise. At high frequencies, the dominant noise will be amplifier voltage noise. In the intermediate region ($\omega \sim \omega_1$) the dominant noise may be microphone noise. In order to pick the optimum signal frequency for a maximum signal-to-noise ratio, it is necessary to compare Eq. (73), which represents the noise, with Eqs. (59) and (64), which give the signal dependence on frequency. Optimization is discussed in the next section.

V. Optimum Design of an Optoacoustic System

In designing an optoacoustic system, the optimum design will depend on the purpose of the system. The equations presented in the preceeding sections of this chapter should be useful in selecting an optimum design. As an example, a common situation will be discussed in this section. Assume that the light source is a laser of fixed beam power W and that the gas container is small enough and the modulation frequency is low enough so that the approximations made in the derivation of Eqs. (59) and (63) are valid. Assuming that the purpose of the system is to measure absorption with the greatest possible sensitivity, then the design goal is to optimize the signal-to-noise ratio. The signal modulation frequency ω, the dimensions of the gas container, and the microphone design are the parameters that can be adjusted.

Assume initially that noise contributed by the amplifier can be neglected. Then, the signal-to-noise ratio can be calculated from Eqs. (59) and (63). Thus, in the absence of amplifier noise, the square of the signal-to-noise ratio is given by

$$\left(\frac{\text{signal}}{\text{noise}}\right)^2 = \frac{(\gamma - 1)^2}{4KT\omega^2}\left(\frac{lA_\text{m}}{V_\text{c}}\right)^2\left(\frac{\omega_\text{m}Q_\text{m}}{K_\text{m}}\right)(\Delta f)^{-1}\alpha^2 W^2. \tag{74}$$

The factor $(\gamma - 1)^2$ can be increased in some circumstances by selecting a monoatomic gas such as helium or argon, rather than nitrogen or air to dilute the sample. This, of course, is not possible in air pollution measurements when the sample is diluted in air and γ is not adjustable. The factor kT in the denominator suggests keeping the system at as low a temperature as possible. Unfortunately, condensation may prevent lowering T very much.

The factor $(lA_\text{m}/V_\text{c})^2$ may be regarded as a coupling coefficient between the microphone and the gas container. It becomes large when a large area microphone and a small cross section container are used. For example, if the gas container is a cylinder of length l and cross-sectional area A_c, then since $V_\text{c} = A_\text{c}l$, the factor $(lA_\text{m}/V_\text{c})^2$ equals $(A_\text{m}/A_\text{c})^2$. This is maximized by choosing the smallest possible cross section area. This area is determined by the requirement to focus the light beam into the container. For example, if the light beam is a laser beam of wavelength λ and lowest-order transverse mode, then the minimum beam diameter at the ends of the gas container is achieved when the beam confocal parameter equals the cell length (Kogelnik and Li, 1966). The resultant beam radius w at the input and output windows is

$$w = \sqrt{\lambda l/\pi}. \tag{75}$$

The beam radius w is the distance from the center of the beam at which the electric field strength is e^{-1} of its value at the beam center. The intensity at this point is e^{-2} of its value at the beam center. Selecting w as the radius of the cylindrical container allows most of the available laser power to enter the container. The container area is

$$A_\text{c} = \lambda l. \tag{76}$$

Substituting the above results into Eq. (74) gives

$$\left(\frac{\text{signal}}{\text{noise}}\right)^2 = \frac{(\gamma - 1)^2}{4KT\omega^2}\left(\frac{A_\text{m}}{\lambda l}\right)^2 \delta^{-1}(\Delta f)^{-1}\alpha^2 W^2. \tag{77}$$

This equation indicates that the signal-to-noise ratio increases as the frequency is reduced. Equation (77) was derived from the assumption that $\omega \gg \tau_T^{-1}$ and thus it is necessary to look at Eq. (58) to understand the low frequency dependence of the signal-to-noise ratio. Equation (58) indicates that the signal-to-noise ratio is maximum when the frequency is very low. Practical considerations such as f^{-1} noise in electronics make it more realistic to chose $\omega \sim 1/\tau_T$, which only reduces the signal-to-noise ratio by a factor of $\sqrt{2}$ from its ideal low-frequency limit.

Amplifier noise was neglected in the previous discussion. When amplifier noise is present, optimum design will depend on the nature of the amplifier noise. The signal amplitude below resonance given by Eq. (59) indicates that the signal amplitude can be increased by choosing a large microphone equivalent volume. The maximum signal amplitude is achieved for $V_m \gg V_c$. In practice, this may be difficult to achieve, and a design goal of $V_m \sim V_c$ approaches within a factor of 2 of the ideal. At low frequencies, the amplifier current noise and the Johnson noise from the bias resistor provide the main source of amplifier noise. If this noise source is larger than the microphone fluctuation noise, then the signal-to-noise ratio is independent of frequency. If the amplifier voltage noise is sufficiently small, then this frequency independence will be maintained up to the resonant frequency ω_1. At this resonance, the signal-to-noise ratio is enhanced by a factor of Q. Thus, it is desirable to operate at resonance if the amplifier voltage noise is small enough and if the dominant noise source is amplifier current noise. Of course, when the dominant noise source is fluctuation noise, then both the signal and the noise are increased equally by the resonance and the optimum operating condition remains $\omega \sim \tau_T^{-1}$. Under the condition where several noise sources contribute or $|V_{na}|^2$ and $|I_{na}|^2$ are not independent of frequency, then the optimum frequency of operation may be different from either τ_T^{-1} or ω_1.

VI. Numerical Example

In order to illustrate the utility of the equations presented in the previous sections, a simple numerical example will be treated in this section. Consider a microphone coupled to a small volume gas container. Let the microphone parameters be given by the values of Table I which correspond to a commercially available 1-in. diameter condenser microphone. Let the light source be a CO_2 laser with emission wavelength of 10.6 μm. Assume also that the cell is a cylinder of volume $V_c = V_m$. Then the radius and length of an optimally designed cell can be calculated from Eqs. (75) and (76):

$$l = \sqrt{V_m/\lambda} = 11.9 \quad \text{cm}, \tag{78}$$

$$w = \sqrt{V_m/\pi l} = 6.33 \times 10^{-2} \quad \text{cm}. \tag{79}$$

TABLE I

Properties of B and K type 4144 Condenser Microphone[a]

m (diaphragm mass) $= 1.40 \times 10^{-5}$ kg
K_m (diaphragm spring constant) $= 3.83 \times 10^{4}$ N m^{-1}
δ (diaphragm damping) $= 9.22 \times 10^{-1}$ N sec m^{-1}
V_m (microphone equivalent volume) $= 0.15$ cm^3
S_m (open circuit sensitivity) $= 5 \times 10^{-2}$ V N^{-1} m^{-2}
C_m (high-frequency capacity) $= 44 \times 10^{-12}$ F
Q_m (mechanical quality factor) $= 0.8$
V_B (bias voltage) $= 200$ V
d (distance between diaphragm and back plate of condenser) $= 2 \times 10^{-5}$ m
A_m (area of microphone diaphragm) $= 1.92 \times 10^{-4}$ m^2
f_m (microphone resonant frequency) $= 8.34$ kHz
ω_m (microphone resonant frequency) $= 5.24 \times 10^{4}$ rad sec^{-1}

[a] Microphones and Microphone Preamplifiers (1975).

The thermal damping time τ_T can be found by equating it to the reciprocal of the frequency at which l_h given by Eq. (34) is equal to the container radius:

$$\tau_T \sim \rho_0 C_P w^2 / 2\kappa. \tag{80}$$

Substituting the values for air at STP [$\rho_0 = 1.29 \times 10^{-3}$ g cm^{-3}, $C_P = 0.24$ cal (g·deg)$^{-1}$, and $\kappa = 5.48 \times 10^{-5}$ cal (cm·sec·deg)$^{-1}$] gives

$$\tau_T \sim 1.13 \times 10^{-2} \quad \text{sec.} \tag{81}$$

The noise equivalent power (NEP) of this detector can be calculated from Eq. (45). Since the effective spring constant is doubled when the microphone is coupled to the gas container, if follows that $\omega_1 Q_1 = 2\omega_m Q_m$. The optimum modulation frequency is $\omega \sim \tau_T^{-1}$. Letting $c = 3.31 \times 10^4$ cm sec^{-1} and $\gamma = 1.403$, Eq. (45) gives

$$\text{NEP} = 3.6 \times 10^{-11} \quad \text{W} (\sqrt{\text{Hz}})^{-1}. \tag{82}$$

Since the absorbed power is $W\alpha l$, it is easy to calculate from Eq. (82) the minimum detectable absorption α_{min}:

$$\alpha_{min} = 3 \times 10^{-12} \quad \text{cm}^{-1} \cdot \text{W} (\sqrt{\text{Hz}})^{-1}. \tag{83}$$

The values of the electrical circuit elements in the equivalent circuit of Fig. 2 can be calculated from Table I. Substituting into Eq. (68) one obtains

$$C^m / C' = 8.7, \tag{84}$$

$$C' = 5.1 \times 10^{-12} \quad \text{F.} \tag{85}$$

When the microphone is coupled to the gas container, the spring constant is doubled and the series capacity C' for the equivalent circuit of the coupled microphone and gas container is one-half the value given by Eq. (85). The signal source is given by substitution into Eq. (65), where V_G is in volts and p_m is in dynes per square centimeter:

$$V_G = 4.4 \times 10^{-2} p_m. \tag{86}$$

The value of the inductance L and resistance R are given by substitution into Eqs. (69) and (70):

$$L = 71 \quad \text{H}, \tag{87}$$

$$R = 4.68 \times 10^6 \quad \Omega. \tag{88}$$

The Johnson noise voltage generated by this resistor is

$$\sqrt{4KTR} = 280 \quad \text{nV} (\sqrt{\text{Hz}})^{-1} \tag{89}$$

Combining Eqs. (89) and (86) one gets the equivalent noise pressure:

$$\sqrt{\overline{p_{\text{noise}}^2}} = 6 \times 10^{-6} \quad \text{dyn}(\text{cm}^2 \cdot \sqrt{\text{Hz}})^{-1}. \tag{90}$$

If the frequency is well below resonance ($\omega \ll \omega_1$), the voltage given by Eq. (90) appears as a voltage at the input terminals of the amplifier:

$$\sqrt{\overline{V_{\text{nm}}^2}} = 29 \quad \text{nV} (\sqrt{\text{Hz}})^{-1}. \tag{91}$$

In order for the amplifier not to be the limiting noise source, the amplifier voltage and current noise must be small compared with this value:

$$\sqrt{|V_{\text{na}}|^2} < 29 \quad \text{nV} (\sqrt{\text{Hz}})^{-1}, \tag{92}$$

$$\sqrt{|I_{\text{na}}|^2} < \omega C_{\text{m}} \sqrt{|V_{\text{nm}}|^2} = 10^{-16} \quad \text{A} (\sqrt{\text{Hz}})^{-1}. \tag{93}$$

Assuming that the current noise is produced by shot noise in the leakage current of the input field effect transistor of the amplifier, this restricts the leakage current I_1:

$$I_1 < 0.06 \times 10^{-12} \quad \text{A}. \tag{94}$$

These results are summarized in Table II.

In the above treatment, only fundamental sources of noise have been discussed. In any real system, additional noise sources may further degrade performance. Some commonly encountered problems are extraneous acoustic noise, vibration-induced noise, and unwanted signal produced by the absorption of the laser beam by the container walls and windows. In selecting an optimum design, these practical problems must be carefully considered.

TABLE II

Summary of Optimum-Design Parameters for an Optoacoustic Detector Using a 10.6-μm CO_2 Laser and a B and K Type 4144 Microphone

l (optical path length in gas) = 11.9 cm
ω (radius of gas container) = 6.33×10^{-2} cm
τ_T(zero order mode damping time) = 1.13×10^{-2} sec
NEF (noise equivalent absorbed power) = 3.6×10^{-11} W $(\sqrt{\text{Hz}})^{-1}$
α_{min} (smallest detectable absorption) = 3×10^{-12} $\text{cm}^{-1} \cdot \text{W} (\sqrt{\text{Hz}})^{-1}$
C' (microphone series resonant capacity) = 5.1×10^{-12} F
V_G/P_m (signal source voltage) = 4.4×10^{-2} V dyn^{-1} cm^{-2}
L (microphone series resonant inductance) = 71 H
R (microphone series resonant resistance) = 4.68 mΩ
$\sqrt{\bar{P}_n^2}$ (equivalent noise pressure at diaphragm) = 6×10^{-6} $\text{dyn}(\text{cm}^2 \cdot \sqrt{\text{Hz}})^{-1}$
$\sqrt{\bar{V}_{\text{nm}}^2}$ (noise voltage at amplifier input resulting from thermal fluctuation in microphone and gas container) = 29 nV $(\sqrt{\text{Hz}})^{-1}$
V_{na}^2 (upper limit on amplifier voltage noise) < 29 nV $(\sqrt{\text{Hz}})^{-1}$
$\sqrt{I_{\text{na}}^2}$ (upper limit on amplifier current noise) < 10^{-16} A $(\sqrt{\text{Hz}})^{-1}$
I_l (upper limit on gate leakage current of amplifier first stage) < 0.06×10^{-12} A

References

Becking, A. G. Th., and Rodenakers, A. (1954). *Acustica* **4**, 96–98.
Dewey, C. F., Jr. (1974). *Opt. Eng.* **13**, 483–488.
Dewey, C. F., Kamm, R. D., and Hackett, C. E. (1973). *Appl. Phys. Lett.* **23**, 633–635.
Goldan, P. D., and Gato, K. (1974). *J. Appl. Phys.* **45**, 4350–4355.
Harshberger, W. R., and Robin, M. B. (1973). *Acc. Chem. Res.* **6**, 329.
Hill, D. W., and Powell, T. (1968). "Non-Dispersive Infra-Red Gas Analysis in Science, Medicine and Industry." Plenum Press, New York.
Kaiser, R. (1959). *Can. J. Phys.* **37**, 1499–1513.
Kerr, E. L., and Atwood, J. G. (1968). *Appl. Opt.* **7**, 915–921.
Kittel, C. (1958). "Elementary Statistical Physics," Part 2. Wiley, New York.
Kogelnik, H., and Li, T. (1966). *Appl. Opt.* **5**, 1550.
Kreuzer, L. B. (1971). *J. Appl. Phys.* **42**, 2934–2943.
Microphones and Microphone Preamplifiers (1975). Brüel and Kjair, DK-2850, Naerum, Denmark.
Morse, P. M. (1948). "Vibration and Sound." McGraw-Hill, New York.
Morse, P. M., and Ingard, K. U. (1961). *In* "Encyclopedia of Physics" (S. Flugge, ed.), Vol. XI/1. Springer-Verlag, Berlin and New York.
Motchenbacher, C. D., and Fitchen, F. C. (1973). "Low Noise Electronic Design." Wiley, New York.
Olson, H. F. (1947). "Elements of Acoustical Engineering," pp. 220–224. Van Nostrand-Reinhold, Princeton, New Jersey.
Rosengren, L.-G. (1975). *Appl. Opt.* **14**, 1960–1976.

2

Energy Transfer Mechanisms

John D. Stettler
Norman M. Witriol

US Army Missile Research, Development, and Engineering Laboratory
Redstone Arsenal, Alabama

I. Introduction

The discovery of the optoacoustic effect, wherein a light beam, modulated in the audio frequencies range and illuminating an absorbant gas, produced a sound wave in that gas was made by Bell (1880, 1881) in 1880, and was immediately investigated by Tyndall (1881) and Röntgen (1881). Thus, we are dealing with a comparatively old field. The phenomenon thus recorded in the 1880s, was however, ignored until 1938 (Viengerov,

1938). Now with the introduction of lasers, where one has a narrow line light source, this method is rapidly becoming a major tool of molecular physics.

Let us consider what causes the sound wave that Bell, Tyndall, and Röntgen first investigated. The gas under investigation is illuminated by a light beam the frequency of which is chosen to be one that is readily absorbed by the gas. Now the question is, "How does this radiation cause sound waves in the gas?" For the purpose of illustrating this phenomenon, we begin by considering the simplest system, namely, a gas consisting of species in which the molecules that can absorb the light have only two energy levels available—the ground state with energy E_0 and the excited state with energy E_1, and the molecules of other molecular species of the gas are always in their ground state.

By absorbing a photon, an absorber molecule in this gas goes from the ground state E_0 to the excited state E_1, the energy difference between the states being $E_1 - E_0 = \hbar\omega_1$, where ω_1 is the angular frequency of the absorbed photon. The molecule can now lose this energy and return to the ground state in the following ways:

1. It can reradiate a photon.
2. It can collide with another molecule of the same species which is in the ground state E_0 and excite it to its excited state E_1.
3. It can collide with any other molecule in the gas and transfer the energy to translational energy, that is, the translational energy of the two molecules after the collision is greater than that before the collision.

The sound wave results from the third process. The increased translational energy in the gas is just increased heat energy. After local thermal equilibrium is reached, this energy has just heated the gas to a higher temperature. If the radiation is modulated at a rate slow compared to the rate of Process 3), the modulation of the light source is seen as modulation in the temperature of the gas. In order to show how this thermal energy is converted to sound waves we assume that the gas can be considered to be an ideal gas. Then $PV = RT$, where P is the pressure of the gas, V is its volume, R is the gas constant, and T the local equilibrium temperature in the gas. Now R and V are constant; therefore, the modulation in T induces a similar modulation in P. Thus, there is a pressure wave with a frequency being that of the modulation frequency of the light beam. But a pressure wave is simply a sound wave. Thus, the modulated light beam produces a sound wave with the same frequency. This phenomenon is what Bell, Tyndall, and Röntgen discovered in 1881, and is what we are applying today.

In Section II we formalize the above concepts and obtain the basic equations for the parameters we measure, namely, the amplitude of and/or the phase lag of the pressure wave in the cell. This section can be omitted by those who are not concerned with the explicit mathematical details, but is included for completeness and for those who wish to know how the assumption of the model, in particular, the number of levels considered, affects the experimentally measured results.

In Section III we discuss what is currently known about the vibration–translation (V–T), vibration–vibration (V–V), translation–translation (T–T), and rotation–rotation (R–R) energy transfer processes. Namely, we shall describe and quantify the physical processes that are important in photoacoustics and determine the rates in the equations derived in Section II.

In Section IV we summarize the current state of the theory.

II. Rate Equation Description

Let us now consider the mathematical formulation used in interpreting the measurements. For simplicity we illustrate the derivation for the simplest model we can consider, the two-level system, and then generalize the results for the $(n + 1)$-level system.

A. Two-Level System

The two-level model is illustrated in Fig. 1. Notationally, each of the parameters k_{ij} represents the nonradiative transition from an initial level i to a final level j [k_{ij} = (the number of collisional energy transfers $(i \to j)$ per unit volume)/(the number of molecules per unit volume × sec)], the detailed evaluation of which we discuss in detail later, and each of the parameters l_{ij},

$$l_{ij} = \rho_\nu B_{ij} + A_{ij}, \tag{1}$$

represents the radiative transitions from an initial level i to a final level j, where ρ_ν is the radiation density at the energy $E_\nu = h\nu$, the energy difference

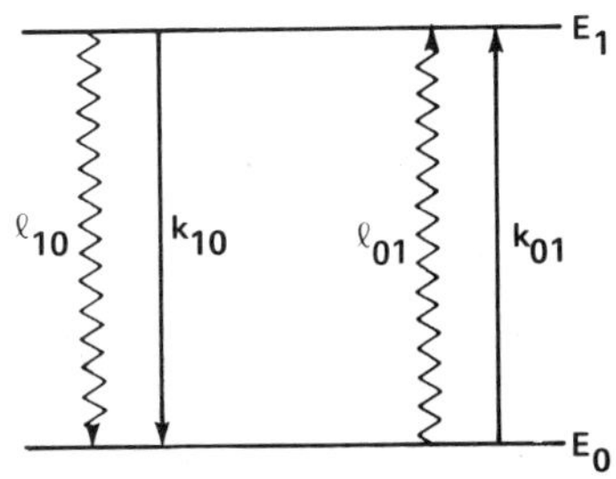

Fig. 1. Schematic representation of the two energy level (E_0, E_1) system showing the radiative and nonradiative transitions.

between levels i and j, B_{ij} ($=B_{ji}$ by detailed balance) are the Einstein coefficients for stimulated emission $i \to j$, and A_{ij} are the Einstein coefficients for spontaneous emission $i \to j$ ($A_{ij} = 0$ if the energy of level $j \geq$ the energy of level i).

Let us consider the time dependence of the number of molecules per unit volume in level i, N_i. The change of N_i with respect to time, $\dot{N}_i$, equals the number of molecules entering level i, minus those leaving level i,

$$\dot{N}_i = (-\text{No. of molecules going out of level } i + \text{No. of molecules going into level } i)/\text{unit time} \tag{2}$$

$$= \left[-\left(\frac{\text{No. going out of level } i/\text{unit time}}{\text{No. currently in } i} \right) \text{No. currently in level } i + \sum_{\text{all levels } j \neq i} \times \left(\frac{\text{No. going into level } i/\text{unit time}}{\text{No. currently in level } j} \right) \text{No. currently in level } j \right]. \tag{3}$$

For the simple two-level system of Fig. 1, Eq. (3) becomes

$$\begin{aligned} \dot{N}_0 &= -(k_{01} + l_{01})N_0 + (k_{10} + l_{10})N_1, \\ \dot{N}_1 &= -(k_{10} + l_{10})N_1 + (k_{01} + l_{01})N_0. \end{aligned} \tag{4}$$

The total number of molecules in all levels is constant; thus

$$N_0 + N_1 = N, \tag{5}$$

where N is the total number of molecules in the system. Since the largest number of molecules is frequently in the ground state, we use Eq. (5) to eliminate N_0. Assuming the pumping of the system is produced by an incident electromagnetic wave modulated at the acoustical frequency ω, we can write the radiation density ρ_v as

$$\rho_v = \rho_v{}^0 + \rho_v{}' e^{i\omega t}. \tag{6}$$

(We note that in actuality we take the real part of ρ_v and in all subsequent expressions the real part is implicitly assumed.) Using Eq. (6), one can rewrite Eq. (1) in the form

$$l_{ij} = l_{ij}^0 + l'_{ij} e^{i\omega t}, \tag{7}$$

where

$$l_{ij}^0 = \rho_v{}^0 B_{ij} + A_{ij}, \qquad l'_{ij} = \rho_v{}' B_{ij} = l'_{ji}. \tag{8}$$

(We note that $l_{ii} = l_{ii}^0 = l'_{ii} = 0$ since this does not represent a transition between two different levels.) Transient terms (i.e., the terms from the initial turning on of the light source) will die out in a time of the order of τ, where

$$1/\tau = (k_{10} + l_{10}^0) + (k_{01} + l_{01}^0). \tag{9}$$

Neglecting these terms (namely, assuming steady state), one can solve an equation of this form by assuming a solution of the form

$$N_1 = \sum_{l=0}^{\infty} N_1{}^l e^{il\omega t}. \tag{10}$$

(We note that if we included an $e^{-i\omega t}$ component in our excitation ρ_v, Eq. (6), we would have to sum over l from $-\infty$ to ∞. This would just add unnecessary complications to our calculation.)

Performing the above operations on Eq. (4) and equating the coefficients of like powers of $e^{il\omega t}$, we can solve Eq. (4) for the $N_1{}^l$, and from Eq. (5), the $N_0{}^l$. As the order m increases, $|N_1{}^m|$ decreases. In most cases, for radiation intensity usually used in optoacoustic spectroscopy, one usually obtains sufficient accuracy by keeping only the zero- and first-order terms. Our solutions for N_1 and N_0 then become

$$N_1 = \frac{N}{[(k_{10}+k_{01})+2\rho_v{}^0 B_{01}+A_{10}]} \times \left\{(k_{01} + \rho_v{}^0 B_{01} + \frac{[(k_{10} - k_{01}) + A_{10}]\rho_v{}' B_{01} e^{i(\omega t - \gamma)}}{\{[(k_{10} + k_{01}) + 2\rho_v{}^0 B_{01} + A_{10}]^2 + \omega^2\}^{1/2}}\right\},$$

$$N_0 = \frac{N}{[(k_{10}+k_{01})+2\rho_v{}^0 B_{01}+A_{10}]} \times \left\{[k_{10}+\rho_v{}^0 B_{01}+A_{10}] - \frac{[(k_{10}-k_{01})+A_{10}]\rho_v{}' B_{01} e^{i(\omega t - \gamma)}}{\{[(k_{10}+k_{01})+2\rho_v{}^0 B_{01}+A_{10}]^2+\omega^2\}^{1/2}}\right\}, \tag{11}$$

where we have set $\tan \gamma = \omega\tau$. Having calculated the level densities, let us now calculate the pressure (acoustic) wave associated with the time-dependent changes in these level densities.

Let us first review why we have an acoustic wave. The incident radiation ρ_v, given by Eq. (6), pumps a significant number of molecules from the ground state 0 to the excited state 1. When the excited molecules decay into the ground state via collisions (the k_{ij} terms), the energy difference between the initial and the final levels, $E_j - E_i = h\nu$, goes into translational energy—that is, the velocities of the colliding molecules increase. Thus we need to find the translational energy per unit volume K. (For simplicity, we will ignore rotational energy.)

The total internal energy of the gas per unit volume is given by

$$U = \left(\sum_{\text{all levels } i} N_i E_i \right) + K. \tag{12}$$

In the two level case, taking the ground state energy $E_0 = 0$,

$$U = N_1 E_1 + K. \tag{13}$$

Thus

$$\dot{U} = \dot{N}_1 E_1 + \dot{K}. \tag{14}$$

By conservation of energy, the change in internal energy is equal to the radiated energy absorbed minus the radiated energy emitted (i.e., the l_{ij} terms). Therefore

$$\dot{U} = (l_{01} N_0 - l_{10} N_1) E_1. \tag{15}$$

Combining Eqs. (14), (15), (4), and (5), we obtain

$$\dot{K} = [(k_{10} + k_{01}) N_1 - k_{01} N] E_1. \tag{16}$$

Now let us consider the relationship between the translational energy per unit volume K and the pressure P. Using thermodynamics

$$dK = (\partial K/\partial T)_V \, dT + (\partial K/\partial V)_T \, dV, \tag{17}$$

where T is the translational temperature and V is the volume. For the cases where the volume is constant, the last term is zero, and we obtain

$$dK = C_V \, dT, \tag{18}$$

where C_V is the specific heat per unit volume at constant volume resulting from translational motion. Integrating this equation, we obtain

$$K = C_V T + f(V), \tag{19}$$

where $f(V)$ is some function that can depend on the volume but is independent of the temperature. Now the equation of state of a real gas can be put into the form

$$P/kT = N + B_2(T) N^2 + \cdots, \tag{20}$$

where the B_n are the virial coefficients, k is the Boltzmann constant, and N is the number of molecules per unit volume. This equation can be rewritten in the form

$$T = (P/kN) - T B_2(T) N - \cdots, \tag{21}$$

$$\equiv (P/kN) a(P, V, N). \tag{22}$$

For an ideal gas $a(P, V, N) = 1$. For a real gas, at sufficiently high temperatures and constant N and V, $a(P, V, N) \simeq$ constant. Substituting Eq. (22) into Eq. (19), we obtain

$$P = (k/C_V a)N[K - f(V)]. \tag{23}$$

Taking the time derivative of Eq. (23) and using Eq. (16) we have

$$\dot{P} = (k/C_V a)N[(k_{10} + k_{01})N_1 - k_{01}N]E_1 \tag{24}$$

Only the oscillating component of P, P_{ose}, is of interest. Thus, substituting Eq. (11) into Eq. (24), retaining only the $e^{i\omega t}$ component, and integrating over t, we obtain

$$P_{\text{osc}} = \frac{kN^2}{C_V a}\frac{E_1}{\omega} \times \frac{(k_{10} + k_{01})[(k_{10} - k_{01}) + A_{10}]\rho_v{}'B_{01}\exp[i(\omega t - \gamma - \pi/2)]}{[(k_{10} + k_{01}) + 2\rho_v{}^0 B_{01} + A_{10}]\{[(k_{10} + k_{01}) + 2\rho_v{}^0 B_{01} + A_{10}]^2 + \omega^2\}^{1/2}} \tag{25}$$

where we have used $i^{-1} = \exp(-i\pi/2)$. Thus, we have the expression for the pressure wave for a two-level system. Let us now consider the generalization of this derivation to the $(n + 1)$-level system.

B. $(n + 1)$-*Level System*

For the purpose of completeness we shall briefly show how the ideas presented in Section A can be extended to the general $(n + 1)$-level system. Using Eq. (3), the $(n + 1)$-level analog of Eq. (4) is

$$\begin{aligned} \dot{N}_0 &= -\left[\sum_{j=0, j\neq 0}^{n} (k_{0j} + l_{0j})\right]N_0 + \left[\sum_{j=0, j\neq 0}^{n} (k_{j0} + l_{j0})N_j\right] \\ &\vdots \\ \dot{N}_n &= -\left[\sum_{j=0, j\neq 0}^{n} (k_{nj} + l_{nj})\right]N_n + \left[\sum_{j=0, j\neq n}^{n} (k_{jn} + l_{jn})N_j\right]. \end{aligned} \tag{26}$$

We set

$$N = \sum_{j=0}^{n} N_j \tag{27}$$

and since the largest number of molecules is frequently in the ground state, as in Section A, use Eq. (27) to eliminate the N_0 equation in Eq. (26), thereby obtaining the set of equations

$$\dot{N}_i = -\left\{\left[\sum_{j=0,j\neq i}^{n} (k_{ij} + l_{ij})\right] + (k_{0i} + l_{0i})\right\} N_i$$

$$+\left\{\sum_{j=1,j\neq i}^{n} [(k_{ji}+l_{ji})-(k_{0i}+l_{0i})]N_j\right\}+(k_{0i}+l_{0i})N, \qquad i=1,\ldots,N. \quad (28)$$

Until now, no assumptions have been made. We shall now assume that we are pumping from the ground state 0 to level m, that is

$$\rho_v = \rho_v{}^0 + \rho_v{}'e^{i\omega t},$$

where ω is the acoustical frequency of the pump and

$$v = (E_m - E_0)/h = \omega/2\pi. \quad (29)$$

We shall further assume that only that transition $(0 \leftrightarrow m)$ is affected by the radiation field ρ_v. Following Eqs. (7) and (8), we expand the radiative transition parameters l_{ij} in the form

$$\begin{aligned} l_{0m} &= l_{0m}^0 + l_{0m}'e^{i\omega t} = \rho_v{}^0 B_{0m} + B_{0m}\rho_v{}'e^{i\omega t}, \\ l_{m0} &= l_{m0}^0 + l_{m0}'e^{i\omega t} = (\rho_v{}^0 B_{0m} + A_{m0}) + B_{0m}\rho_v{}'e^{i\omega t}, \\ l_{ij} &= l_{ij}^0 = A_{ij} + \bar{\rho}_{ij}B_{ij}, \qquad i > j \\ &= \bar{\rho}_{ij}B_{ij}, \qquad j \leq i \end{aligned} \left\{\begin{array}{l}\text{for all the} \\ \text{other } l_{ij}\text{'s,}\end{array}\right. \quad (30)$$

where $\bar{\rho}_{ij}$ is the black body radiation at the energy $|E_i - E_j|$. For mathematical simplification, we define the following parameters:

$$\begin{aligned} a_i &= \left(\sum_{j=0,j\neq i}^{n} (k_{ij} + l_{ij}^0)\right) + (k_{0i} + l_{0i}^0), \qquad b_i = (l_{i0}' + l_{0i}')\delta_{im}, \\ c_{ij} &= -[(k_{ji} + l_{ji}^0)] + (k_{0i} + l_{0i}^0), \quad d_i = l_{0m}'\,\delta_{im}, \\ e_i &= (k_{0i} + l_{0i}^0), \qquad f_i = l_{0m}'\,\delta_{im}, \end{aligned} \quad (31)$$

where δ_{im} is the Kronecker delta function

$$\begin{aligned} \delta_{im} &= 1, \qquad i = m \\ &= 0, \qquad i \neq m. \end{aligned}$$

Equation (28) can then be written in the form

$$|\dot{N} = -(|A + e^{i\omega t}|B)|N + (|E + e^{i\omega t}|F)N, \quad (32)$$

where

$$|N = \begin{pmatrix} N_1 \\ N_2 \\ \vdots \\ N_{(n-1)} \\ N_n \end{pmatrix}, \qquad |A = \begin{pmatrix} a_1 & c_{12} & \cdots & & c_{1n} \\ c_{21} & a_2 & \cdots & & c_{2n} \\ \vdots & & & & \vdots \\ & & & a_{(n-1)} & c_{(n-1)n} \\ c_{n1} & c_{n2} & \cdots & c_{n(n-1)} & a_n \end{pmatrix},$$

$$|B = \begin{pmatrix} b_1 & d_1 & \cdots & & d_1 \\ d_2 & b_2 & \cdots & & d_2 \\ \vdots & & \vdots & & \vdots \\ d_{(n-1)} & & \cdots & b_{(n-1)} & d_{(n-1)} \\ d_n & & \cdots & d_n & b_n \end{pmatrix}, \qquad |E = \begin{pmatrix} e_1 \\ e_2 \\ \vdots \\ e_{(n-1)} \\ e_n \end{pmatrix}, \qquad |F = \begin{pmatrix} f_1 \\ f_2 \\ \vdots \\ f_{(n-1)} \\ f_n \end{pmatrix}. \tag{33}$$

As in Section A, we assume a solution of the form

$$N_i = \sum_{l=0}^{\infty} N_i^l e^{il\omega t}, \qquad |N = \sum_{l=0}^{\infty} |N^l e^{il\omega t}. \tag{34}$$

(We note that since Eq. (28) is a linear equation in the N_i, if there were r pumping terms with different ω's, the solution N_i would be taken as a sum over all r terms of the form of Eq. (34). In addition the radiative parameters l_{ij} would have to be specified in a manner analogous to Eq. (30).) Substituting Eq. (34) into Eq. (32), we obtain the tensor equation

$$\sum_{l=0}^{\infty} [|A + i\omega l]|N^l e^{il\omega t} = -\sum_{l=1}^{\infty} |B|N_i^{l-1} e^{il\omega t} + [|E + e^{i\omega t}|F]N. \tag{35}$$

Equating powers of $e^{i\omega t}$, we thereby obtain the equations

$$\begin{aligned} |A|N^0 &= |EN \\ (|A + i\omega)|N^1 &= -|B|N^0 + |FN \\ (|A + i\omega l)|N^l &= -|B|N^{l-1}, \qquad l \geq 2. \end{aligned} \tag{36}$$

These equations are readily solved via Cramer's rule. The $N_0{}^l$ terms are then simply obtained from Eq. (27). As in the two-level case, as the order l increases, the $|N_i{}^m|$ terms generally decrease, and one can usually obtain sufficient accuracy by retaining only the zero- and first-order terms in Eq. (36).

Let us now consider the acoustic wave. The internal energy, taking $E_0 = 0$, is given by

$$U = \sum_{i=1}^{n} N_i E_i + K, \tag{37}$$

where K is the translational energy of the system. Thus

$$\dot{U} = \sum_{i=1}^{n} \dot{N}_i E_i + \dot{K} \tag{38}$$

and by conservation of energy, as in Eq. (15),

$$\dot{U} = \sum_{i,j=0}^{n} l_{ij} N_i (E_j - E_i). \tag{39}$$

Combining Eqs. (38), (39), and (26) we obtain

$$\dot{K} = \sum_{i,j=0}^{n} (k_{ij} N_i - k_{ji} N_j) E_i, \tag{40}$$

where we have taken $k_{ii} = 0$. As in Eq. (24), the time derivative of the pressure of the gas is given by

$$\dot{P} = (k/C_V a) N \dot{K}. \tag{41}$$

Substituting Eq. (40) into Eq. (41), using the solutions to Eq. (36) and retaining only the $l = 1$ solution thereof, and intergrating over time, we find that the oscillatory component P_{osc} of the pressure P becomes

$$P_{\text{osc}} = \frac{1}{\omega} \frac{k}{C_V a} N \left[\sum_{i,j=0}^{n} (k_{ij} N_i^{\,1} - k_{ji} N_j^{\,1}) E_i \right] \exp\left[i \left(\omega t - \frac{\pi}{2} \right) \right]. \tag{42}$$

Thus, we have a general formalism for an arbitrary $(n + 1)$-level system. Let us now consider the application of this procedure.

In applying this procedure to a specific example the first consideration is the number of levels needed in the model. Inclusion of unnecessary levels makes the mathematics tedious, whereas inclusion of too few levels leads to erroneous results, as the experimental data will not fit the theoretical curve. A good experimental example of using too few levels is given by Cannemeijer *et al.* (1971). In general, for interpreting experimental data, one must carefully consider each specific system individually. One must make a model, including all the major processes taking place in that system, and use Eq. (42) or a generalization thereof for interpreting the data.

III. Energy Transfer Processes

In this section we show how to convert between the experimentally determined rate constants k_{ij} and the theoretically derived probability of energy transfer per elastic collision P_{ij}, discuss the magnitude of the energies involved in the various energy transfer processes, and review the current theories of these processes.

A. *Experimentally Determined versus Theoretically Derived Parameters*

In Section II we used the parameter k_{ij} = [the number of collisional energy transfers ($i \to j$) per unit volume]/(the number of molecules per unit volume × sec). However, in theoretical computations one computes P_{ij} = probability that in a single gas kinetic collision one obtains the energy transfer $i \to j$. [Note, for example, that the general form of P_{ij} for vibrational processes is $P(\sum_\alpha A(\alpha(i \to j)), \sum_\beta B(\beta(k \to l)))$ = probability that in a collision between molecules A and B, the modes α of A go from vibrational levels $\alpha(i)$ to vibrational levels $\alpha(j)$, while the modes β of B go from vibrational levels $\beta(k)$ to vibrational levels $\beta(l)$.] Generally, the current theories consider transitions between only one mode per molecule at a time. To compare theory with experiment we therefore need to know the conversion between P_{ij} and k_{ij}.

From their definitions

$$k_{ij} = \bar{Z}_{AB} P_{ij}, \tag{43}$$

where $\bar{Z}_{AB}$ is the number of gas kinetic collisions per unit volume per second between the respective molecules, which we will take as A and B. The quantity $\bar{Z}_{AB}$ generally used (see, for example, Taylor and Bitterman, 1969; Herzfeld and Litovitz, 1959) is based on assuming (*i*) only binary collisions are important, and (*ii*) the population of the levels of the modes are given by a Boltzmann distribution. Under these assumptions one obtains (Taylor and Bitterman, 1969)

$$\begin{aligned} \bar{Z}_{AB} &= (8\pi kT/\mu_{AB})^{1/2} \sigma_{AB}^2 [1 - \exp(-\hbar\omega/kT)] \qquad \text{(for V–T transfer)} \\ &= (8\pi kT/\mu_{AB})^{1/2} \sigma_{AB}^2 \qquad \text{(for V–V transfer)}, \end{aligned} \tag{44}$$

where μ_{AB} is the reduced mass, (mass A + mass B)/(mass A) × (mass B), σ_{AB} is the diameter of the collision cross section, generally (under the assumption that the molecules act as billiard balls) taken (Hirschfelder *et al.*, 1954) as

$$\sigma_{AB} = \tfrac{1}{2}(\sigma_A + \sigma_B), \tag{45}$$

where σ_A and σ_B are respectively the hard-core radii of the Leonard–Jones potentials for molecules A and B, and ω is the angular frequency of the vibrational energy lost in the V–T collision. Thus, one now has the means of converting between experimental data and theoretical computation.

B. *Order of Magnitudes of Energies Transferred*

Let us now consider the magnitude of the energies involved in energy transfer processes. Energy transfer rates depend on the amount of energy transferred as well as the process effecting the transfer. Thus, we first give estimates of the energies associated with the various internal motions of molecules. Electronic energies E_{el} are of the order of

$$E_{el} \approx \hbar^2/ma^2, \tag{46}$$

where a is a measure of the size of the molecule, m is the mass of the electron, and $\hbar$ is Planck's constant divided by 2π. Thus, frequencies associated with changes in electronic energy are in the ultraviolet to visible spectrum. Molecular binding energies involved in vibration have the form $\frac{1}{2}KX^2$, where X is the amount that the molecule has stretched and K is the effective spring constant. In order to stretch a molecule so that $X \approx a$ an amount of energy has to be imparted to it of the order of its dissociation energy, which is governed by electronic energies. Thus we can estimate that $Ka^2 \approx E_{el}$, which leads to the vibrational energy spacings E_{vib} being of the order of

$$E_{vib} \approx \hbar\sqrt{K/M} \approx (\sqrt{m/M})E_{el}, \tag{47}$$

where M is the molecular mass. Since $m/M \approx 10^{-4}$ for most molecules, this shows that vibrational frequencies are in the infrared to far infrared. Rotational energy spacings, E_{Rot} are

$$E_{Rot} \approx \hbar^2/I \approx \hbar^2/Ma^2 \approx (m/M)E_{el}, \tag{48}$$

where I is the molecular I, moment of inertia. Except for hydrides, these frequencies are generally in the far-infrared to microwave part of the spectrum; for the hydrides these are in the infrared. The final type of molecular motion is translational. This provides a reservoir into which energy from the other motions can flow. The energy per molecule is not governed by intrinsic molecular parameters but is given by

$$E_{trans} \approx \tfrac{3}{2}KT, \tag{49}$$

where K is Boltzmann's constant and T is the ambient absolute temperature. Except at cryogenic temperatures this energy is greater than molecular rotational spacings. For low-frequency vibrations or at elevated temperatures it can also be comparable to vibrational energy spacings. Generally it is not the ratio of E_{trans} to energy transferred in a process that is important. Rather it is the ratio of E_{trans} to the energy transferred into or out of translational motion that is significant. For transfer of a quantum of vibration totally into translational motion (a V–T process) these two ratios are the same. However, consider the transfer of a quantum of vibration of energy $\hbar\omega_1$ from one molecule to another molecule whose vibrational quantum is $\hbar\omega_2$. This exchange (a V–V process) involves the exchange of $\hbar\omega_1$ energy between molecules and the release to or absorption from translation of an energy $\hbar|\omega_1 - \omega_2|$. For resonant cases, for example, CO_2 and N_2, the latter quantity can be quite small. Thus, for V–V processes the two ratios can be significantly different. It is also important to realize that the latter ratio, energy to or from translation to average translational energy, is the quantity that governs the ratio of the forward to back process rate via the Boltzmann factor.

C. *Physics of Energy Transfer Mechanisms*

Processes that effect the transfer of energy from a molecule are necessarily nonadiabatic (Herzfeld and Litovitz, 1959; Landau and Teller, 1936). In this context an adiabatic process is one in which parameters that determine the energy are changed so slowly that quantum numbers do not change. For example, when two molecules collide, their intermolecular distance changes with time. This makes the intermolecular force change with time, which induces a time-varying force on the internal vibrational motion. If all the changes occur slowly compared with a vibrational period, the vibrational motion will adjust *adiabatically* to the changes so that the vibrational quantum state will not change. In such a collision no V–T process will result. Thus, the essence of energy transfer processes is the breakdown of adiabaticity. Anything that will affect this will impact the magnitude of the rate. In order to clearly identify the factors that control adiabaticity we consider a V–T process in detail.

In Fig. 2 we show the simplest possible collision in which atom A collides with atom B along a line coincident with the axis of the molecule BC. We assume that there are only two forces involved in this collision. The first is the spring that couples C and B and represents the internal vibrational motion of the molecule BC. The other force is the intermolecular force between BC and A. Although two extremes can be identified when one or the other of these two forces is much larger than the other, this alone does not determine the degree of adiabaticity of the collision. From the previous discussion it can be seen that the rate at which A approaches BC is also crucial.

First, we consider the limiting case in which the atoms are hard elastic spheres. The intermolecular force is very short range, being nonzero only

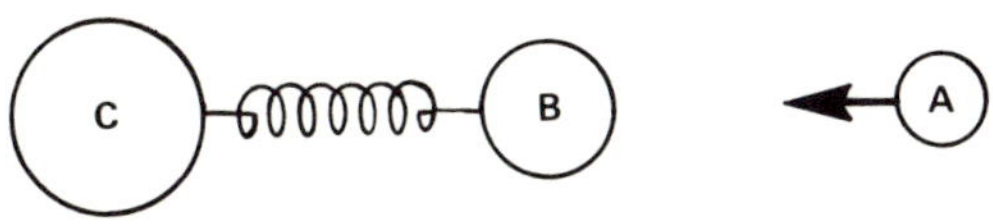

Fig. 2. Collinear collision between an atom A and a diatomic molecule BC.

when A physically touches B. If, in this case, we also take the spring to be weak relative to the impulsive force of A on B, then we can neglect the forces exerted by the spring on both B and C. The collision between A and B is just an elastic collision between spheres. By conservation of energy and momentum their velocities can be calculated from the knowledge of the original velocities of A and B. If B is assumed to be initially at rest, the collision will be the inverse to V–T relaxation. If B is initially moving, either vibration excitation or deexcitation can occur. In any event after the collision B will be traveling toward C, compressing the spring, whereas A will be traveling away from CB. So by the time B has been stopped by the spring and rebounded back to its original position, A has moved sufficiently far enough away so that no secondary collision can occur. Thus, the most energy allowed by conservation condition is transferred in this situation; it is the extreme nonadiabatic limit. Because C does not participate at all, it plays the role of a spectator and this is the spectator limit (Levine and Bernstein, 1974). The fractional amount of the energy of A transferred to vibration of BC is governed by the kinematic factor $m_A m_C/m_B(m_A + m_B + m_C)$, which must be small for the spectator limit to be valid (Secrest and Johnson, 1966).

The opposite case, the adiabatic limit of weak energy transfer, encompasses several different situations. The spring BC can be so strong that BC responds as a unit when A strikes B. The force between B and A can be long range and relatively weak so that at every instant during the collision all atoms are essentially in kinematic equilibrium. Accelerations are always very small. As A approaches BC, the kinetic energy of A is turned into potential energy of repulsion between A and B. Then as A retreats this energy is returned to A as kinetic energy. No net energy is left in vibration. Finally the kinetic energy of A can be significantly less than the quantum of vibration of *BC* so that quantum restrictions mitigate against energy transfer.

The above qualitative considerations can be subsumed into the statement that the important ratio is that of the time of interaction or collision t_c to the period of vibration t_v. If $l = vt_c$ is a characteristic length of interaction for a relative velocity v, then for an assumed repulsive exponential potential the probability of energy transfer per gas kinetic collision varies as (Landau and Teller, 1936)

$$P \sim \exp -(\omega l/v), \tag{50}$$

where ω is the vibrational angular frequency. From an average over a thermal distribution the effective velocity for Eq. (50) turns out to be

$$v_{\text{eff}} = (\omega l k T/\bar{m})^{1/3}, \tag{51}$$

where the effective mass

$$\bar{m} = m_A(m_B + m_C)/(m_A + m_B + m_C) \tag{52}$$

is the reduced mass for the BC, A system. Thus, for relaxation processes mediated by strong, short-range potentials the functional variation with physical quantities for the energy transfer rate is

$$P \sim \exp\left\{-\left(\frac{m_A(m_B + m_C)}{(m_A + m_B + m_C)} \frac{(l\Delta E)^2}{\hbar^2 kT}\right)\right\}^{1/3}, \tag{53}$$

where ΔE is the energy transferred, l is the interaction length (typically of the order of 0.2 Å), m_A is the mass of the relaxation inducing atom/molecule, m_B is the mass of an atom on the molecule being relaxed that is involved in the internal mode being relaxed, and m_C is the mass of the rest of the molecule being relaxed. The multiplicative constant to make Eq. (53) an equality is generally not given accurately by theory but must be obtained by comparison with experiment. An illustration (Cannemeijer and DeVries, 1973) of the mass factor relationship is given in Fig. 3.

Although Eq. (53) was formulated in the context of V–T processes, it also holds for any process for which nonresonant, short-range forces are the mediating agent. Thus, the process described could also be the transfer of electronic energy to translational energy. In this case ΔE is not the separation of the electronic levels in energy at infinite intermolecular distance but rather the energy difference at the intermolecular distance at which the transfer takes place. Because energy levels of the electronic states can be strongly perturbed when molecules collide, these two energy separations

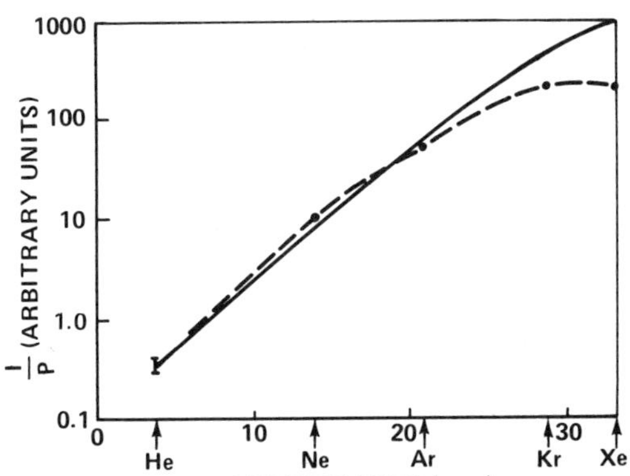

Fig. 3. P^{-1} (CO_2 – X) for deexcitation of the (0, 1, 0) level of CO_2 by noble gases versus reduced mass. The solid line is obtained from the S.S.H. theory (Herzfeld and Litovitz, 1959).

can be quite different. Illustrative V–T and V–V rates are given in Tables I and II, respectively.

Equation (50) could also hold for rotational relaxation, where ΔE is the splitting of rotational energy levels, or for intermolecular vibrational energy

TABLE I

Illustrative V–T Rates (Room Temperature)

System	P	Ref.
O_2–O_2	4.8×10^{-8}	Herzfeld and Litovitz (1959)
N_2–N_2	9.1×10^{-9}	Herzfeld and Litovitz (1959)
CO_2–CO_2	4.6×10^{-5}	Herzfeld and Litovitz (1959)
Cl_2–Cl_2	2.9×10^{-5}	Herzfeld and Litovitz (1959)
N_2O–N_2O	1.2×10^{-4}	Herzfeld and Litovitz (1959)
COS–COS	1.0×10^{-4}	Herzfeld and Litovitz (1959)
CS_2–CS_2	1.25×10^{-4}	Herzfeld and Litovitz (1959)
HCl–HCl	1.27×10^{-4}	Chen (1971)
HBr–HBr	1.23×10^{-4}	Chen (1971)
O_2–H_2	5.0×10^{-5}	Herzfeld and Litovitz (1959)
Cl_2–He	1.1×10^{-3}	Herzfeld and Litovitz (1959)
Cl_2–N_2	2.3×10^{-5}	Herzfeld and Litovitz (1959)
HCl–He	1.8×10^{-7}	Zittel and Moore (1973)
HCl–CO	3.4×10^{-4}	Chen and Moore (1971)
HBr–HCl	2.32×10^{-4}	Chen (1971)
HBr–He	$\leq 7.5 \times 10^{-7}$	Chen (1971)
HBr–Ne	$\leq 7.5 \times 10^{-7}$	Chen (1971)
HBr–Ar	$\leq 2.3 \times 10^{-7}$	Chen (1971)

TABLE II

Illustrative V–V Rates (Room Temperature)[a]

System	P	ΔE (cm^{-1})
CO–O_2	2.2×10^{-7}	587
N_2–O_2	1.0×10^{-6}	775
N_2–CO	5.1×10^{-5}	188
NO–CO	1.0×10^{-4}	267
CO_2–CO	6.6×10^{-4}	206
N_2O–N_2	3.9×10^{-2}	107
HCl–N_2	1.1×10^{-4}	555
HCl–CO	3.4×10^{-4}	743
HCl–CH_4	9.4×10^{-3}	30
HCl–HBr	6.0×10^{-3}	327

[a] From Moore (1973).

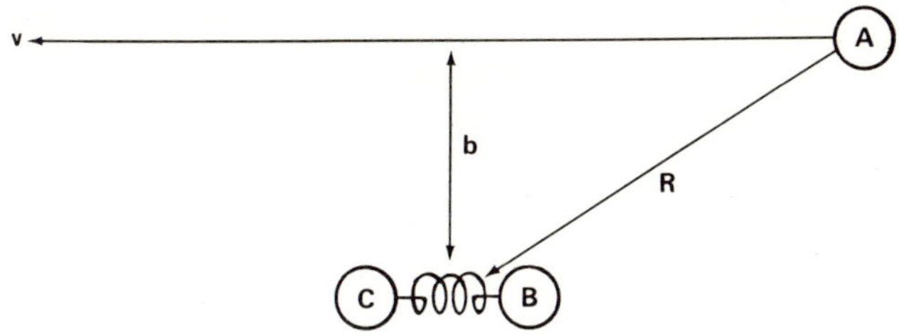

Fig. 4. Nonzero impact parameter b collision between an atom A and a diatomic BC.

transfer, where ΔE is the net energy transferred to translation, not the vibrational energy transferred between molecules. In these two cases ΔE can be quite small. Resonant effects can become important. Long-range electromagnetic multipole–multipole interactions can dominate the energy transfer process. The picture based on Fig. 1 is no longer valid because grazing, nonzero impact parameter collisions such as those in Fig. 4 can become important (Sharma and Brau, 1967). The dominant collisions have $\omega l/v \sim 1$ where l is now a function of b. After an integral over b and an average over a thermal velocity distribution, the energy transfer probability becomes

$$P \sim (\bar{m}/kT)(Q_A Q_B/\hbar\sigma^n)^2, \tag{54}$$

independent of ΔE. Here Q_A and Q_B are the dominant multipole moments of A and B, respectively, σ is the minimum distance between molecular centers, $\bar{m}$ is again the reduced mass, and the inverse distance dependence of the multipole–multipole interaction involved is distance to the nth power (for dipole–dipole $n = 2$, dipole–quadupole $n = 3$, etc). Note the difference in temperature dependence between Eqs. (53) and (54). Rotational relaxation of polar molecules is governed by Eq. (54) (Cross and Gordon, 1966; Oka, 1974). Also resonant V–V processes (Sharma and Brau, 1967; Moore, 1973) such as D_2–HCl and N_2–CO_2 (asymmetric stretch) follow Eq. (54) for temperature less than $10^{3\circ}$K. Although multipole–multipole forces exist in the H_2–HCl system, its V–V transfer rate should follow Eq. (53) because of the large energy defect ΔE that must be given to translation. In all V–V processes it is important to note that ΔE is the sum of a vibrational energy mismatch and a rotational energy mismatch. Since these can have different signs a V–V process can be made resonant by the inclusion of rotational energy transfer (Sharma, 1969). For highly polar molecules like HF, their relaxation by polar molecules can actually be a V–R–T process (Ahl and Cool, 1973).

In summary, we present Table III where the various energy modes are listed as well as the general order of magnitude for the probability per collision of energy transfer taking place. Note that for R–R the probability

TABLE III

Comparative Order of Magnitudes for the Probability per Collision of Various Energy Transfer Processes

Process	Resonant		Nonresonant
V–V	0.2–0.05	—	0.05–0.001
V–R	0.1–0.01	—	0.01–0.001
V–T	—	0.01–0.0001	—
R–R	10–0.2	—	0.2–0.05
R–T	—	0.1–0.01	—

may be greater than one. This is possible (Cross and Gordon, 1966) since, because of long-range forces, the inelastic cross section can be larger than the elastic cross section.

IV. Conclusion

In this chapter we have presented the reader with the methodology to make and interpret a model of an optoacoustic experiment along with a brief review of the current theory of energy transfer mechanisms. For further details on the latter, the interested reader can use the following reviews as guides to delve into the specialist literature. For a general tutorial introduction into intermolecular energy transfer (including electronic) the reviews of Levine and Bernstein (1974) and of Nikitin (1974) are a good start. The theory of rotation and vibration energy transfer is reviewed by Secrest (1973) and of V–T in a more expository manner by Rapp and Kassel (1969). Rotational energy transfer, both theoretical and experimental, is treated by Oka (1974) as well as by Toennies (1974), who also considers vibrational energy transfer. Vibrational energy transfer is also surveyed by Moore (1973), who mostly considers diatomic molecules, and by Flynn (1974), who also considers large molecules. Specific coverage of laser studies of energy transfer in small molecules is given by Weitz and Flynn (1975).

References

Ahl, J. L., and Cool, T. A. (1973). *J. Chem. Phys.* **58**, 5540.

Bell, A. G. (1880). *Proc. Am. Assoc. Adv. Sci.* **29**, 115; (1881). *Philos. Mag.* **11**, 510.

Cannemeijer, F., De Vascuncelos, M. H., and De Vries, A. E. (1971). *Physica (The Hague)* **53**, 77.

Cannemeijer, F. and De Vries, A. E. (1973). *Physica (The Hague)* **70**, 135.

Chen, H.-L. (1971). *J. Chem. Phys.* **55**, 5551.
Chen, H.-L., and Moore, C. B. (1971). *J. Chem. Phys.* **54**, 4080.
Cross, R. J., Jr., and Gordon, R. G. (1966). *J. Chem. Phys.* **45**, 3571.
Flynn, G. W. (1974). *In* "Lasers in Chemistry" (C. B. Moore, ed.), Chapter 6. Academic Press, New York.
Herzfeld, K. F., and Litovitz, T. A. (1959). "Absorption and Dispersion of Ultrasonic Waves," pp. 260–348. Academic Press, New York.
Hirschfelder, J. O., Curtis, C. F., and Bird, R. B. (1954). "Molecular Theory of Gases and Liquids," pp. 1110–1111. Wiley, New York.
Landau, L. and Teller, E. (1936). *Phys. Z. Sowjetunion* **10**, 34.
Levine, R. D., and Bernstein, R. B. (1974). "Molecular Reaction Dynamics." Oxford Univ. Press, London and New York.
Moore, C. B. (1973). *Adv. Chem. Phys.* **23**, 41.
Nikitin, E. E. (1974). *In* "Physical Chemistry" (W. Jost, ed.), Vol. VIA. Academic Press, New York.
Oka, T. (1974). *Adv. At. Mol. Phys.* **9**, 127.
Rapp, D., and Kassal, T. (1969). *Chem. Rev.* **69**, 61.
Röntgen, W. C. (1881). *Philos. Mag.* **11**, 308.
Secrest, D. (1973). *Annu. Rev. Phys. Chem.* **24**, 379.
Secrest, D., and Johnson, B. R. (1966). *J. Chem. Phys.* **45**, 4556.
Sharma, R. D. (1969). *Phys. Rev.* **177**, 102.
Sharma, R. D., and Brau, C. A. (1967). *Phys. Rev. Lett.* **19**, 1273.
Taylor, R. L., and Bitterman, S. (1969). *Rev. Mod. Phys.* **41**, 26.
Toennies, J. P. (1974). *Chem. Soc. Rev.* **3**, 407.
Tyndall J. (1881). *Proc. R. Soc. London* **31**, 307, 478.
Viengerov, M. L. (1938). *Dokl. Akad. Nauk. SSSR* **19**, 687.
Weitz, E., and Flynn, G. (1975). *Annu. Rev. Phys. Chem.* **25**, 275.
Zittel, P. F., and Moore, C. B. (1973). *J. Chem. Phys.* **58**, 2922.

3

Design of Optoacoustic Systems

C. Forbes Dewey, Jr.

Department of Mechanical Engineering
Massachusetts Institute of Technology
Cambridge, Massachusetts

I. Introduction

The optoacoustic effect has been employed in a wide variety of experimental situations, each having its particular needs with respect to sensitivity, accuracy, dynamic range, sample conditioning, and measurement environment. In this chapter we review the capabilities and limitations of various optoacoustic system designs, and present quantitative relations for predicting system performance.

The pressure change in an enclosed gas volume caused by the absorption of infrared radiation has been used since about 1947 in the design of commercial gas-filled radiation detectors. Pressure changes are detected by a capacitance microphone in a manner similar to that described by Kreuzer in Chapter 1. Golay, Luft, Sall, Venigrov, Delaney, Powell, and others have contributed to the development and optimization of these detectors; their work is summarized in the excellent book by Hill and Powell (1968). These pneumatic detectors are designed to measure the amount of radiation that has passed through a separate sample cell. The radiation transmitted with and without a sample gas present can be used to measure the concentration of an absorbing species present in the sample, as in a nondispersive infrared (NDIR) analyzer. Pneumatic detectors of this type are designed to absorb a sizable fraction of the incident radiation within the spectral absorption bands of the fill gas.

Current interest in spectrophones arises primarily from the increased sensitivity that may be obtained by the use of lasers as sources of optical excitation (Kerr and Atwood, 1968; Kerr, 1973; Kreuzer, 1971, 1974). The spectrophone has become a method of choice for measuring atmospheric pollutants at the ppb level and in measuring gas absorption coefficients α as low as $10^{-8}\ \mathrm{cm}^{-1}$. The optical power absorbed by the gas sample, and hence the acoustic signal, is proportional to σn, where σ is the absorption cross section (cm^2) and n is the number density of absorbers (cm^{-3}). Two different modes of operation are therefore possible: the first is when σ is known for the absorbing species, so that the acoustic signal is proportional to the concentration n; the second mode is when the absorbing species concentration is known, and it is desired to determine the absorption cross section. In the latter case, absorption spectra may be recorded as the wavelength of the incident light source is varied.

A substantial body of optoacoustic literature is concerned with understanding collisional and radiative energy transfer processes in gases (cf. Bates *et al.*, 1970; Evans *et al.*, 1972; Grabiner *et al.*, 1972; Tripodi, 1970; Tripodi and Vincenti, 1971; Woodmansee and Decius, 1962; and the reviews by Bauer, 1972; Delaney, 1959; Kaiser, 1959; and Read, 1967–68). The energy transfer mechanisms are described in more detail by Kreuzer, Chapter 1 and Stettler and Witriol, Chapter 2. This discussion is restricted to situations in which local thermodynamic equilibrium exists at each point within the sample chamber, and the pressure rise in the gas may be computed by applying conservation of energy and the constitutive equations of state for the sample gas.

In Section II we discuss sample chamber design, citing different geometrical and optical configurations that have been investigated or appear promising. Acoustic signals are calculated, and scaling with sample chamber size and geometry is discussed. A simple theory is presented to estimate background signals arising from window heating. Spurious signals from aerosols, electronic noise, and other sources are estimated. Consideration is given to sampling system design, including the adverse effects of turbulent flow noise. In Section III amplitude and wavelength modulation of the excitation are discussed in a general way. Operation at an acoustic resonance is analyzed, and theoretical and experimental results for acoustic loss mechanisms are summarized. A general treatment of wavelength-modulated light illustrates certain significant advantages that may be obtained if a broadly tunable laser optical source is used. Section IV provides a comparison between optoacoustic spectroscopy and other analytic techniques for spectroscopic purposes and for detecting trace atmospheric constituents.

II. Sample Chamber Design

A portion of the modulated optical energy incident on the gas sample being analyzed is absorbed, and this produces a modulated pressure signal within the sample chamber. A microphone placed in the chamber wall records the magnitude of this induced pressure change. Many geometric and optical arrangements for the sample chamber, optical source, and microphone can be conceived. In this section, we discuss the advantages and limitations of some of these various designs. Particular attention will be given to acoustically resonant systems (Dewey *et al.*, 1973).

A. Calculation of Acoustic Signals

A typical optoacoustic system is illustrated in Fig. 1 of Chapter 1. The incident light source is modulated sinusoidally at a frequency f, yielding a modulated energy input into the gas. If the fractional absorption by the gas is small, then the pressure modulation amplitude is linearly proportional to the gas absorption coefficient α (cm^{-1}), and the incident light intensity, I (ergs $cm^{-2}sec^{-1}$), but inversely proportional to the sample chamber cross-sectional area, A_c (cm^2).

In calculating the amplitude of the induced acoustic signal, it is convenient to decompose the signal into a linear combination of orthogonal modes p_j, each of which corresponds to a normal acoustic mode of the

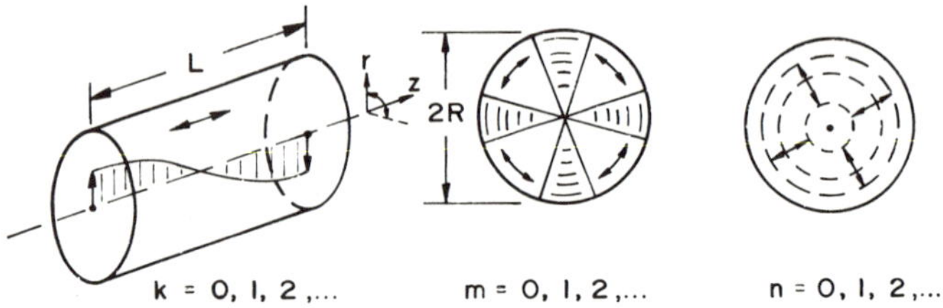

Fig. 1. Longitudinal, azimuthal, and radial modes of a cylindrical sample chamber. (From Dewey *et al.*, 1973).

sample chamber. For a cylindrical chamber of length L and radius R, these normal modes are given by (e.g., Morse and Ingard, 1968; Compton, 1969)

$$\begin{aligned} p_j(r, \phi, z, t) &= P_j \cos(m\phi)\cos(k\pi z/L) J_m(\alpha_{m,n}\pi r/R)\exp(-i\omega_j t), \\ \omega_j &= \pi c_0[(k/L)^2 + (\alpha_{m,n}/R)^2]^{1/2}, \end{aligned} \tag{1}$$

where the eigenvalues k, m, and n define the longitudinal, azimuthal, and radial modes of the system as depicted in Fig. 1. Here $\alpha_{m,n}$ is the nth solution of the equation $(dJ_m/dr) = 0$ at $r = R$, and c_0 is the speed of sound in the sample gas. The term P_j is a normalization coefficient such that

$$V_c^{-1} \int p_j^* p_k \, dV = \delta_{jk}, \tag{2}$$

with p_j^* being the complex conjugate of p_j and δ_{ij} equal to unity for $k = j$ and zero otherwise. V_c is the sample chamber volume (cm^3). The spatial variation of pressure within the sample chamber is then

$$p(\omega, \mathbf{r}) = \sum_j A_j p_j(\mathbf{r}, \omega_j), \tag{3}$$

where the amplitude of each orthonormal mode is A_j and may be determined according to the relations given in Chapter 1. Similar forms for the mode amplitude calculations have been given by Rosengren (1973, 1975) and Kamm (1976).

From Chapter 1, Eq. (24), the amplitude A_j (dynes cm^{-2}) of the jth normal pressure mode caused by an energy absorption rate per unit volume $H(\omega, \mathbf{r})$ modulated at a frequency ω ($\omega = 2\pi f$, sec^{-1}) and with a spatial variation throughout the sample chamber volume is given by

$$A_j = -\frac{i\omega(\gamma - 1)(V_c)^{-1} \int p_j^*(\mathbf{r}, \omega_j) H(\mathbf{r}, \omega) \, dV}{\omega^2[1 - (\omega/\omega_j)^2 - i(\omega/\omega_j Q_j)]}, \tag{4}$$

where γ is the ratio of specific heats (5/3 for a monatomic gas, 7/5 for a diatomic gas), Q_j is the acoustic resonant enhancement of the jth mode at frequency ω_j, and the volume integral extends over the total volume of the sample chamber.

The most common geometrical configuration for optoacoustic sample cells is a cylinder, with the excitation beam traveling parallel to the tube axis and being radially symmetric about it. If, in addition, the excitation beam suffers negligible attenuation along the axis, then the energy deposition rate per unit volume, $H(\mathbf{r}, \omega)$, reduces to $\alpha I(r, \omega)$, where α is the absorption coefficient (cm^{-1}) and $I(r, \omega)$ represents the radial variation of the excitation beam intensity [ergs $(cm^2\text{-sec})^{-1}$]. The orthogonal normal pressure modes $p_j(r, \omega_j)$ and eigenfrequencies ω_j are then

$$p_j(r, \omega_j) = P_j J_0(\pi\alpha_{0,j} r/R) \exp(-i\omega t). \qquad \omega_j = (\pi c_0 \alpha_{0,j}/R), \qquad P_j = [J_0(\pi\alpha_{0,j})]^{-1}. \tag{5}$$

No axial or azimuthal pressure modulations are produced because the excitation beam has no variation with ϕ or z. These conditions apply to all further discussion in the chapter.

The amplitudes A_j depend on the coupling between the beam intensity profile, $I(r, \omega)$ and the normal modes $p_j(r, \omega_j)$ by virtue of the integral in Eq. (4). To illustrate this coupling more explicitly, one may express $I(r, \omega)$ using the same orthonormal set of eigenfunctions as that used to represent the induced pressure*:

$$I(r, \omega) = \sum_n B_n h_n(r), \qquad h_n(r) = J_0(\pi\alpha_{0,n} r/R)[J_0(\pi\alpha_{0,n})]^{-1} e^{-i\omega t}. \tag{6}$$

Then the coupling integral contains terms of the form

$$(V_c)^{-1} \int p_j^* h_n \, dV = \delta_{jn}.$$

Thus, only the component B_j of the excitation beam will produce a normal pressure mode with a spatial and temporal form given by $p_j(r, \omega_j)$.

Kamm (1976) has used this normalization to investigate the effects of excitation beam diameter on the coupling of the absorbed energy to the acoustic field. As a representative beam profile, he used the Gaussian form

$$I(r, \omega) = G \exp[-(r/a)^2] \exp(-i\omega t).$$

If the characteristic width a of the beam is less than half of the chamber radius R, the algebra simplifies to yield the following expression for the

* The beam intensity $I(r, \omega)$ is represented as a sinusoidal function of time. Inasmuch as the heat deposited in the gas by the excitation beam is always positive, the appropriate value of $I(r, \omega)$ to be used in these calculations is *one-half* the peak optical intensity measured by a power meter. Thus, half the energy deposited in the gas contributes to a steady-state temperature distribution that has no ac component. Similarly, in Eqs. (7) and (8), P_{abs} is equal to *one half* the peak power absorbed by the gas.

BIRMINGHAM UNIVERSITY LIBRARY

induced pressure:

$$p(r,t) = \sum_j \left(\frac{-i\omega}{\omega_j^{\,2}}\right) \frac{\gamma - 1}{(1 - \omega^2/\omega_j^{\,2} - i\omega/\omega_j Q_j)} \frac{P_{\rm abs}}{V_{\rm c}} \frac{\exp(-\mu_j)}{J_0^{\,2}(\pi\alpha_{0,j})} \times J_0(\pi\alpha_{0,j} r/R) \exp(-i\omega_j t). \tag{7}$$

In this equation, $P_{\rm abs}$ is one half the peak power absorbed by the gas and

$$\mu_j = (a/R)^2 (2\pi\alpha_{0,j})^{-2}.$$

For example, the amplitude of the lowest-order radial pressure mode ($j = 1$) excited at the resonant frequency $\omega = \omega_1$ is

$$A_1 = \frac{(\gamma - 1)Q_1}{\omega_1} \left(\frac{P_{\rm abs}}{V_{\rm c}}\right) \frac{\exp(-\mu_1)}{J_0^{\,2}(\pi\alpha_{0,1})} J_0(\pi\alpha_{0,1} r/R). \tag{8}$$

The variation of A_1 with the excitation beam radius appears only in the term μ_1; A_1 is a maximum for $(a/R) \to 0$ and is 40% of the maximum for $(a/R) = 1/2$. One concludes from these calculations that the amplitude of the induced pressure is not particularly sensitive to the details of the excitation beam shape as long as a strong overlap exists between the beam profile and the spatial amplitude of the mode being excited.

Equation (7) illustrates that there are two competing effects that determine the amplitude of the pressure response. The first is the dependence on the excitation frequency ω; a substantial enhancement occurs when ω is close to one of the resonant frequencies ω_j, and $A_j \sim Q_j/\omega_j$. Clearly, operation at a low-frequency resonance is desired. The second effect is contained in the term $P_{\rm abs}/V_{\rm c}$; maximum pressures are obtained for a small chamber volume. For a cylindrical chamber of radius R, $\omega_j \sim R^{-1}$ and $V_{\rm c} \sim R^2$ and a maximum response is achieved, for the same absorbed power, with the smallest practical chamber radius consistent with a good spatial overlap between the excitation beam and the excited pressure mode.

It is useful to compare the response of optoacoustic systems operated with uniform radial excitation ($j = 0$, $\omega_0 = 0$) and at the lowest-order radial resonance ($j = 1$, $\omega_1 = (\pi\alpha_{0,1} c_0/R)$). As Kreuzer (1971; Chapter 1) has shown, the term Q_0/ω_0 becomes the thermal time constant τ_T for $j = 0$ so that

$$A_0 = \frac{i(\gamma - 1)}{\omega} \left(\frac{P_{\rm abs}}{V_{\rm c}}\right) \left(1 + \frac{i}{\omega\tau_T}\right)^{-1}.$$

The thermal time constant τ_T is approximately (Kreuzer, 1971)

$$\tau_T \approx R^2/2.5\delta,$$

where δ is the thermal diffusivity $(k/\rho C_p)$ of the gas and is approximately $0.3\ \mathrm{cm^2\ sec^{-1}}$ for air at STP. The magnitude of the induced pressure at the external cylindrical surface is

$$|p(\omega)| = \left|\frac{i(\gamma - 1)}{\omega}\left(\frac{P_{\mathrm{abs}}}{V_c}\right)\left(1 + \frac{i}{\omega\tau_T}\right)^{-1}\right|,$$

and the equivalent result for excitation at the lowest-order resonance ω_1 is given by Eq. (8), with $r = R$. Taking $\mu_1 \ll 1$ gives

$$p(\omega_1) = \frac{(\gamma - 1)Q_1}{\omega_1}\left(\frac{P_{\mathrm{abs}}}{V_c}\right)[J_0(\pi\alpha_{0,1})]^{-1}.$$

Practical considerations of beam alignment and amplifier noise suggest that operating frequencies $f = (\omega/2\pi)$ should exceed 20 Hz and chamber diameters should be larger than 5 mm. Under these conditions, $\tau_T \gtrsim 10^{-1}$ sec and $\omega\tau_T \gg 1$. The uniform excitation, or zero-order, response is then

$$|p(\omega)| = [(\gamma - 1)/\omega](P_{\mathrm{abs}}/V_c),$$

and the ratio of the pressure response for equal values of P_{abs}/V_c in the two operation modes is

$$p(\omega_1)/p(\omega) = (\omega/\omega_1)Q_1/J_0(\pi\alpha_{0,1}). \tag{9}$$

In deriving Eq. (9), it has been assumed that (P_{abs}/V_c) is the same for both resonant and nonresonant conditions. With laser sources, it is possible to focus the incident light to very small beams, so that small volume cells may be used with nearly equal P_{abs}. In practice, it is difficult to couple small diameter cells to commercial microphones which range from 8 to 25 mm in diameter. Kreuzer (1971) and Bruce (1976) have shown that large-area microphones which cover the entire outer circumference of a cylindrical cell may be fabricated from aluminized Mylar; this method may be used to produce custom assemblies of small diameter that serve as both a sample cell and as a microphone. With 25-mm diameter commerical microphones, practical minimum cell diameters are on the order of 25–50 mm.

The acoustic Q of resonant systems can be seriously degraded by geometrical designs that do not have smooth boundaries conforming to the desired normalized resonant mode shape p_j. With cylindrical cells of 50- and 100-mm diameter, and microphone diameters up to 25 mm, Kamm (1973), Goldan and Goto (1974), and Max and Rosengren (1974) have achieved values of Q_1 at resonance for the first-order radial acoustic mode of 770–890. These experimental results suggest that substantial resonant enhancement can be achieved in practice. The resonant frequency ω_1 is proportional to the speed of sound $c_0 = (\gamma RT/M)^{1/2}$, and for small changes in mean molecular weight M, temperature T, and specific heat ratio γ

$$\delta\omega_1/\omega_1 = \tfrac{1}{2}(\delta\gamma/\gamma + \delta T/T - \delta M/M). \tag{10}$$

From Eq. (4), the behavior of the amplitude A_1 near resonance is

$$\left|\frac{A_1(\omega)}{A_1(\omega_1)}\right| \approx \left[1 + Q_1{}^2\left(1 - \frac{\omega^2}{\omega_1{}^2}\right)^2\right]^{-1/2}. \tag{11}$$

The acoustic power in the standing wave is proportional to $A_1{}^2(\omega)$, so that the half-power points correspond to values of $\omega \equiv \omega_1 \pm \delta\omega$ for which

$$\delta\omega/\omega_1 = (2Q_1)^{-1}. \tag{12}$$

The practical conclusion is that close temperature and frequency control are required to maintain a highly tuned system near resonance. For a room-temperature system, a 1°C change at constant frequency in a detector with $Q_1 = 500$ would decrease the pressure aplitude A_1 by about a factor of 2. However, it is a relatively straightforward matter to incorporate detection circuitry that adjusts ω to maximize the microphone signal and compensate for slow changes in temperature and molecular weight. We conclude that values of Q_1 on the order of 300–500 may be achieved in practical circumstances.

Returning to Eq. (9), we may assess the relative merits of operating in a resonant or a nonresonant mode with a fixed cylindrical sample chamber. Nonresonant operation is favored in systems using a low frequency and small chamber diameter. Choosing a minimum practical frequency of $f = \omega/2\pi =$ 20 Hz and a diameter of 2.54 cm with air at STP as the test gas gives $f_1 =$ $(\omega_1/2\pi) =$ 16,600 Hz, and taking $Q_1 = 500$, Eq. (9) predicts $p(\omega_1)/p(\omega) = 1.5$. For smaller chamber diameters, nonresonant operation is preferred.

Schnell and Fischer (1975a, b) employed an 8-cm diameter cylindrical cell for which $f_1 = 5050$ Hz. Operation at f_1 yielded about 20 times the sensitivity obtained at 100 Hz. Equation (9) predicts that their system had a value of Q_1 of about 400, a value not inconsistent with other results.

B. Sample Chamber Design

Figure 2 illustrates a number of optical configurations which have been utilized or proposed for optoacoustic sample cells. Single-pass resonant and nonresonant configurations have been discussed in the preceding paragraphs. Resonant multiple-pass operation has been explored by Kamm (1973) and Goldan and Goto (1974).

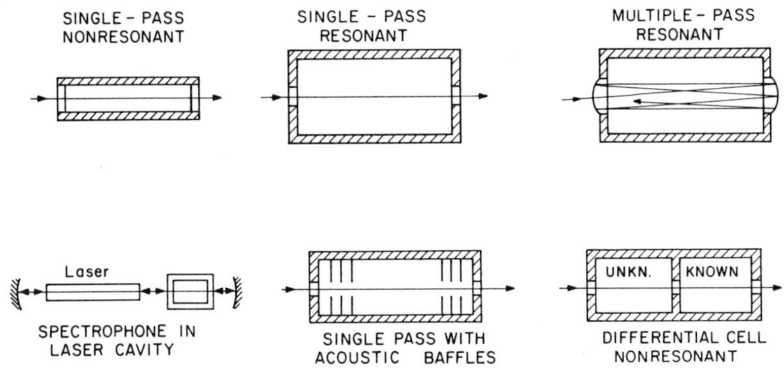

Fig. 2. Schematic drawings of various designs for optoacoustic cells.

Placing the spectrophone within the laser cavity is a logical extension of the use of intracavity absorption cells in gas lasers. An intracavity absorption cell was employed by Chakerian and Weisbach (1973), who measured changes in the gain of a laser operated near threshold. Inasmuch as the maximum advantages of this configuration are realized for low-gain lasers with very low output coupling, serious problems would arise in reducing reflection losses at the cell windows.

One of the major limitations to the sensitivity of spectrophones is the background signals that arise from window heating. A single pass system with acoustic baffles separating the central test section from the regions of disturbance is shown in Fig. 2. In resonant systems, this can lead to a substantial reduction in Q_1 (Dewey, 1974; Bruce *et al.*, 1976). More direct methods for reducing the background signals arising from window heating are discussed in Sections III and IV.

A method which was used by Deaton *et al.* (1975) to balance out background signals is also illustrated in Fig. 2. Two spectrophone cells of identical characteristics are placed in series in the path of the optical source. The differential signal between the cells is related to the absorption in the measuring cell. If both cells contain nonabsorbing gas, then only background signals are produced and the differential signal should be zero. Leaving the reference cell void of absorbing gas and subtracting its signal from that of the sample cell should produce a net signal free of background. However, one can achieve equally accurate background subtraction by direct comparison of the sample cell signals with and without absorbers present. Further, any accumulation with time of scatterers or absorbing film on the sample cell windows will lead to different background signals in the two cells.

The differential cell method does have utility as a simple calibration standard when the spectrophone is being used to measure the concentration or absorption characteristics of the sample gas. A known concentration of absorbing species is placed in the calibration cell and the calibration signal is then proportional to the incident beam intensity and the known concentration in the calibration cell. If $(\alpha L) \ll 1$ and background signals are negligible, the ratio of the sample and calibration signals is directly proportional to the concentration ratio of the two cells. Identical geometries and good thermal contact between the cells will also compensate for resonant frequency shifts arising from thermal drift.

One of the most promising designs for practical application is a multiple-pass resonant cylindrical chamber (Kamm and Dewey, 1973; Goldan and Goto, 1974). Figure 3 is an illustration of a favorable configuration of such a cell (Kamm, 1976). By placing the reflective surfaces external to the acoustic cavity, any absorption in the reflecting films will not contribute to the acoustic background of the system. Following a suggestion of Rosengren (1975), a favorable method of introducing the test gas is to provide inlet and outlet ports at the node of the resonant acoustic mode. By the "principle of reciprocity" (see Morse and Ingard, 1968, p. 312) pressure disturbances at a node of a resonance mode will not contribute to that mode, but radial velocity fluctuations will since the perturbation velocity is a maximum at the pressure node. The lowest-order radial mode of a cylinder has a node at $(r/R) = 0.63$.

The multiple-path arrangement of Fig. 3 provides a radially symmetric excitation function with a mild axial variation. The beam is concentrated near the axis, providing good coupling to the lowest-order radial resonance. With antireflection coatings on the inner surfaces of the end windows, it should be straightforward in practice to achieve an enhancement of 20 in the effective intensity of the excitation source. This geometry is also well suited to the use of blackbody radiation sources, which could be attractive for inexpensive pollution detection instruments (Dewey, 1974). An alternative multiple-pass arrangement that requires a laser source is described by Goldan and Goto (1974).

Other system configurations not illustrated in Fig. 2 have been suggested by Rosengren (1975) and Claspy *et al.* (1976).

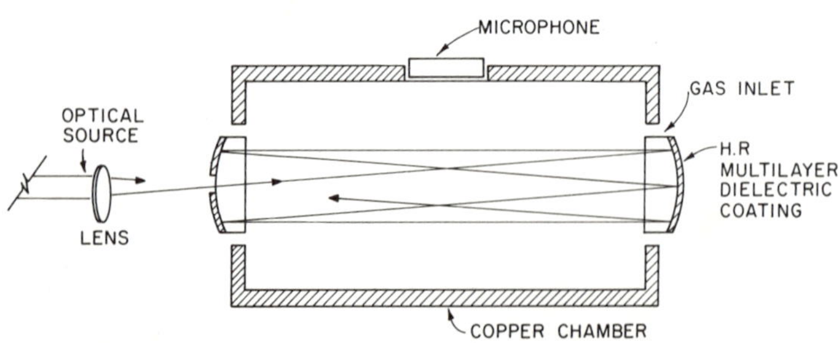

Fig. 3. Multiple-pass configuration utilizing external mirrors.

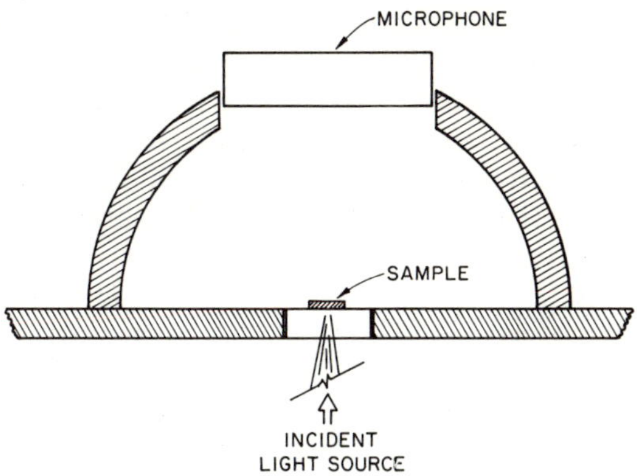

Fig. 4. Optoacoustic chamber for analyzing solid samples or exploring thin film absorption coefficients. Note that a cylindrical or spherical chamber may be operated at an acoustic resonance.

While the use of various sample cell configurations with gaseous absorbers has been emphasized, these same techniques are also useful with solid samples, as discussed by Rosencwaig in Chapter 9 of this book. Figure 4 shows a particularly favorable design for exploring the absorption spectra of thin films placed on a solid substrate. Incident light from a laser or monochrometer is incident on the sample, and the modulated temperature rise of the film produces an acoustic response in the surrounding gas which is detected by the microphone. As suggested by Fig. 4, the chamber can be cylindrical or spherical, allowing operation at an appropriate resonant frequency. For highly absorbing samples with large thermal inertia, it is desirable to illuminate the side of the sample which is interior to the chamber. This apparatus would be useful, for example, in characterizing the absorption of thin films of dielectric material on coated optics. A somewhat different design was employed by Kerr (1973) to measure surface-film fractional absorptions in the 10^{-4}–10^{-5} range.

C. *Practical Design Considerations*

There have been a large number of investigations within the past six years that have used optoacoustic methods to measure gaseous absorption coefficients and to detect trace atmospheric constituents. The emphasis of these studies has been to achieve high sensitivity, large dynamic range, and reproducible, quantitative accuracy. A number of impediments to achieving these goals have arisen, and it is useful to summarize these effects here and to indicate practical solutions which have been proposed.

The most persistent difficulty encountered in achieving high sensitivity has been the background signal that arises from absorption of the excitation beam on the sample cell windows and radiation scattered to the chamber walls. This background signal is proportional to excitation intensity, so with high-power laser sources this effect can easily dominate the sensitivity limit. For example, Kreuzer *et al.* (1972) had a background signal 125 times larger than the noise signal from the microphone and amplifier.

The dominant source of background noise is absorption at the cell windows, and this effect will now be described in more detail. A similar analysis has been presented by Parker (1973) and recast by Rosencwaig and Gersho (1975). From Parker's calculations, it is clear that the dominant source of acoustic disturbance is a thin film of absorbing material near the surface, rather than bulk absorption. In the case of highly transparent windows with light incident from either the gas or the solid side, the absorption region is sufficiently thin to allow the assumption that the heating takes place in a very small region at the gas–solid interface which has negligible thermal capacitance. Figure 5 depicts this situation.

The heat flux into the film $q_0(t)$ must equal the sum of the heat conducted into the solid $q_s(t)$ plus the heat conducted into the gas $q_v(t)$. These heat fluxes are given by the Fourier law

$$q_0(t) = q_v(t) + q_s(t), \tag{13}$$

$$q_v(t) = -k_v(\partial T_v/\partial y_+)_{y_+=0} \equiv -k_v(\partial T/\partial y_+)_0, \tag{14}$$

$$q_s(t) = -k_s(\partial T_s/\partial y_-)_{y_-=0} \equiv -k_s(\partial T/\partial y_-)_0. \tag{15}$$

Further, $T_s = T_v$ at $y = 0$.

We assume a sinusoidal heat input, and express its magnitude as

$$q_0(t) = q_{max} \sin(\omega t - \phi), \tag{16}$$

where ϕ is an arbitrary phase angle. The temperature gradients in Eqs. (14) and (15) must also vary sinusoidally with time, and with the same phase since T is continuous across the gas-solid interface. This boundary-value problem corresponds to Stokes' second problem (often called the Rayleigh problem; see, e.g., Schlichting, 1968); the temperature distributions within

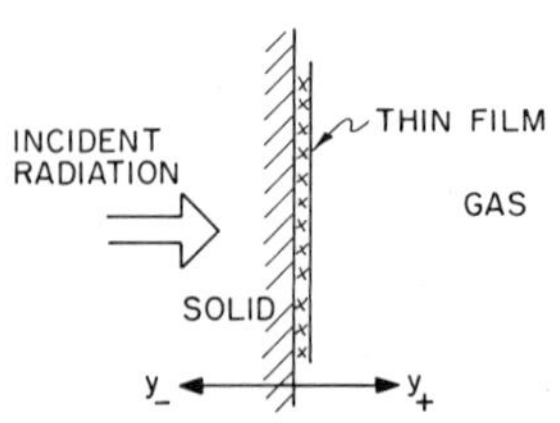

Fig. 5. Absorption model for window heating problem.

the two domains will be governed by the same heat conduction equation,

$$\partial T/\partial t = (k/\rho C_p)\,\partial^2 T/\partial y^2, \tag{17}$$

and the temperature T, thermal conductivity k, density ρ, and specific heat C_p take on the subscripts s and v for the solid and the gas, respectively. Equation (17) is subject to the boundary condition

$$(T)_{y=0} = T_0 \cos \omega t. \tag{18}$$

It is assumed that the thermal boundary layers in the two regions are sufficiently thin that both regions appear semiinfinite; that is, one seeks bounded solutions for which $T \to 0$ as $y_{\pm} \to \infty$.

The solution satisfying Eqs. (17) and (18) is

$$T(y,t) = T_0 e^{-\alpha^* y} \cos(\omega t - \alpha^* y), \qquad \alpha^* = (\omega \rho C_p/2k)^{1/2} \tag{19}$$

with y measured positive away from the interface into each region. Differentiating Eq. (19) with respect to y and evaluating the gradient at $y = 0$, we find

$$(\partial T/\partial y)_0 = 2^{1/2} \alpha^* T_0 \sin(\omega t - \pi/4). \tag{20}$$

Equation (20), with appropriate subscripts, may be used in Eqs. (13)–(16) to determine the heat flow into the solid and the gas as a function of the amplitude of the sinusoidal heat input to the film, $q_{\max}$. The phase angle ϕ between the temperature oscillation and the heat input is $(\pi/4)$, and the maximum temperature excursion at the interface is

$$T_0 = q_{\max}/2^{1/2}(\alpha_v^* k_v + \alpha_s^* k_s). \tag{21}$$

The temperature distributions in the gas and the solid are shown schematically in Fig. 6.

The sinusoidal heat input into the gas produces acoustic disturbances at the same frequency as the pressure waves produced by gas absorption; as such, it is a source of background signal. To minimize this signal, one must maximize the fraction of q_0 which is conducted into the solid. The

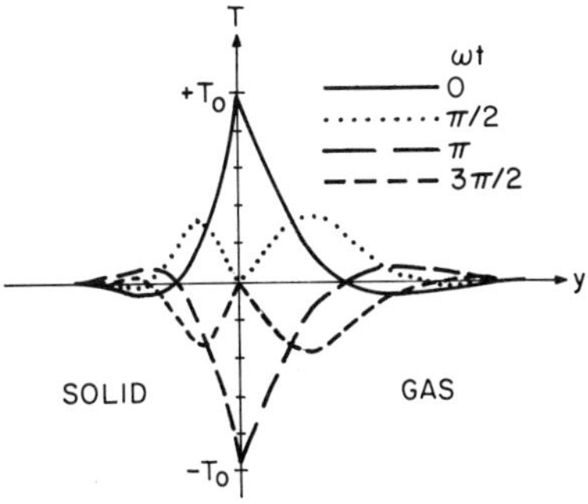

Fig. 6. Temperature distributions in the gas and solid on either side of an absorbing film.

ratio (q_v/q_s) is

$$q_v/q_s = (\rho_v C_{p_v} k_v / \rho_s C_{p_s} k_s)^{1/2}. \tag{22}$$

For transparent substrates, there is a definite "pecking order" of desirability for use as window materials. The value of $(\rho_s C_{p_s} k_s)^{1/2}$ for sapphire, for example, is 6 times that of quartz and 23 times that of plastic. For solid substrates, for example, chamber walls that receive scattered radiation, obvious choices (in order of decreasing preference) are copper, silver, gold, and aluminum.

The above analysis has assumed that the thermal wave in the gas produced by surface absorption is confined to a narrow boundary layer near the surface, and the results are thus restricted to the operating condition $\omega\tau_T \gg 1$. As was pointed out earlier, this is the usual regime for practical cell dimensions; it also includes the case of acoustically resonant operation.

The magnitude of the induced background pressure can be determined, in the resonant case, by treating the heat input to the gas as a *negative* thermal loss. For further details of these calculations, see Kamm (1976) and Kreuzer, Chapter 1.

There are two distinct types of surface heating which must be distinguished for a cylindrical sample cell having an excitation beam propagating along its axis. The first is window heating which, because the heating occurs along the axis, provides an in-phase background contribution to the pressure signal. The second is caused by the radiation scattered by aerosols and dust; the scattered radiation would be absorbed primarily by surface layers on the cylindrical walls. Aerosol scattering can be significant; an empirical approximation to the aerosol scattering coefficient at sea level in the infrared region of the spectrum is

$$\sigma\,(\text{cm}^{-1}) = 1.3 \times 10^{-6}[\lambda\,(\mu\text{m})]^{-0.422},$$

which, in the spectral "windows" of the atmosphere, is comparable to the absorption coefficients of trace atmospheric constituents.

The lowest-order radial mode, $J_0(\pi\alpha_{0,1}r/R)$, is a pressure wave that has opposite sign at the axis (where primary heating by the excitation beam occurs) and at the outer wall (where the microphone is normally located). For this reason, surface absorption on the cylindrical walls contributes a pressure signal out of phase with respect to the primary signal. The total background from these two sources, window heating and wall heating, may either be in phase or out of phase with respect to the desired signal, depending upon the relative magnitudes of the two sources. This is illustrated in Fig. 7.

In practice, the effects of aerosol scattering are generally smaller than those produced by window heating. Figure 8, from the work of Kamm (1973, 1976), illustrates that the departure from linearity at low signal levels is well described by a constant background in phase with the primary excitation.

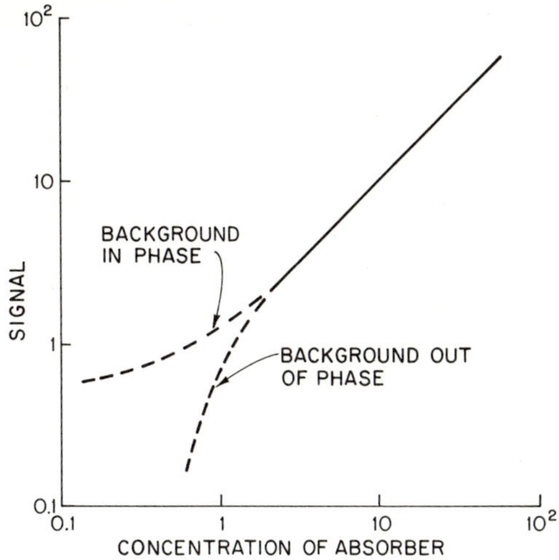

Fig. 7. Nonlinearity of acoustic response in the presence of background signals.

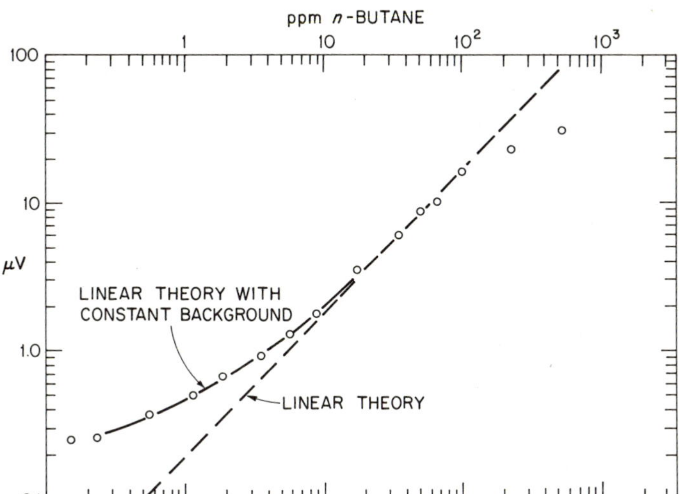

Fig. 8. Typical results for a resonant acoustic chamber with in-phase background signal from window heating. (From Kamm, 1976.)

The effects of window heating can be minimized by increasing the chamber length. For a fixed excitation frequency, the background should be inversely proportional to length. The most successful method of eliminating this background is to use a wavelength-tunable laser and modulate the wavelength of excitation. Surface absorption is often independent of wavelength so that no ac signal would be produced with a constant intensity input. The gas absorption would be modulated as the excitation wavelength changes relative to the center of the gas absorption line. This technique has been used very successfully by Patel (1973) and Patel *et al.* (1974) to reduce the window heating background. Further discussion of wavelength modulation is given later in this chapter (Section III.C).

There are numerous sources of additional background noise in spectrophones. One is external noise, which may arise from ambient acoustic and building vibrations and, in the case of mechanically chopped optical sources, from the chopper itself. Use of a phase-sensitive lock-in amplifier and a long integration time is successful in discriminating against ambient vibration, but chopper noise cannot be eliminated by this method. If the flow of sample gas into the chamber is turbulent, very strong background signals are observed. A useful artifice for minimizing the effects of flow-induced pressure fluctuations is to introduce the sample gas at a node of the standing wave of a resonant mode (see Fig. 3).

Mechanical forces produced by an imbalance in the chopper wheel or by application and relaxation of electromagnetic forces on the laser components during modulation can also be transmitted to the microphone. Further, careful shielding to avoid electromagnetic pickup by the microphone and amplifier is an absolute necessity.

In using spectrophones for trace gas analysis, sample conditioning becomes a major concern. The species being detected may be adsorbed on sample walls (Schnell and Fischer, 1975b). With changing concentration, the measured acoustic amplitude is representative of the concentration being irradiated and not necessarily either the volumetric average concentration within the cell or the concentration entering the cell at the time of measurement. If rapidly changing concentrations are expected, gas inlet and exit ports should be suitably placed to promote mixing of the incoming sample and the gas resident in the chamber.

A problem common to all air pollution measurement systems is the preparation of accurate calibration samples. Permeation tube systems are frequently employed to provide ppb–ppm standards. (See, e.g., the references cited in the review article by Saltzman and Cuddeback, 1975.)

III. Wavelength and Amplitude Modulation of the Excitation Source

Excitation beam modulation is required to produce the acoustic signals detected in a spectrophone. This modulation may be of amplitude or of wavelength, or a combination of the two. As has been demonstrated previously in this chapter, the frequency with which the modulation is accomplished changes the pressure amplitude as well as the magnitude of background signals that are present.

This section begins with a brief morphology of modulation techniques, indicating some of the potential advantages and disadvantages of each. Next, resonant systems are considered and estimates are made of the various loss mechanisms which contribute to the calculation of Q_j. This is followed by an extensive treatment of the effects of wavelength modulation and the use of broadly tunable sources for absorption line spectroscopy.

A. Methods of Excitation Beam Modulation

The following list of beam modulation methods is indicative of the numerous possible modes of spectrophone operation:

1. Low-frequency nonresonant amplitude modulation, fixed wavelength. This mode of operation is simple and particularly appropriate for continuous wave (CW) fixed-frequency infrared gas lasers. A disadvantage is the existence of high background signals. Useful for quantitating absorption of high-power laser sources by various trace atmospheric constituents. Equivalent results are obtained using a low-repetition rate pulsed laser.

2. High-frequency resonant amplitude modulation, fixed wavelength. Requires sample cell diameter at least twice beam diameter. Resonant frequency inversely proportional to diameter. Useful in environments where low-frequency noise sources are present. Some advantage over Method 1 in reducing background from window heating and aerosol scattering. Multiple-pass operation facilitated.

3. Low-frequency amplitude modulation with slow wavelength tuning (Kreuzer and Patel, 1971). Very useful for spectroscopic studies, as with a tunable laser diode.

4. CW source with small modulation of source wavelength. Provides derivative signal that discriminates against background signals. Typical sources are magnetically and thermally tuned semiconductor lasers, etalon-tuned dye lasers, pressure-broadened gas lasers, and spin-flip Raman lasers.

5. Source CW in wavelength and amplitude, but absorption line tuned over small range using Zeeman or Stark effect. Provides derivate signal. Very useful for Zeeman-active gases (Bonczyk and Ultee, 1972; Kaldor *et al.*, 1972; Freund and Sweger, 1975). As yet, this method has not been utilized to produce optoacoustic modulation.

6. Source amplitude constant, but wavelength modulated substantially. As discussed in Section III.C, new and useful information can be obtained with this mode of operation.

7. Single-pulse excitation of resonant cell. The decay of the acoustic wave produced by a pulsed laser gives information on the cavity Q (Kamm, 1976) and the absorption coefficient of the sample gas (Bruce *et al.*, 1976).

Methods 2 and 4–6 may be used with modulation frequencies which correspond to acoustic resonances of the sample cell. The calculation of the enhancement factors Q_i is now discussed.

B. Acoustically Resonant Operation

The most common geometry for resonant spectrophones is a cylindrical chamber with an optical excitation source propagating along the axis (Dewey *et al.*, 1973). To achieve good coupling between the excitation and the standing wave, the beam intensity should be confined to the vicinity of the axis (see the discussion following Eq. 11). Operation of the lowest-order radial mode has produced resonant enhancement factors Q_1 as large as 800.

Wave dissipation in an acoustic cavity arises from volumetric effects (as in the free propagation of sound waves) and in surface effects. In small spectrophones operated at 0.1 atm or higher pressure, the dominant energy losses involve surface effects. They are:

(1) wave reflection losses resulting from the compliance of chamber walls;
(2) dissipation at the microphone diaphragm, which, for a properly designed microphone, is equal to the electrical signal energy produced;
(3) viscous and thermal losses occurring in the gas boundary layer near the chamber surface;
(4) losses resulting from wave scattering at surface obstructions, for example, gas inlet ports and microphone mounts.

In many practical circumstances, viscous and thermal losses at the wall are the primary effects that determine Q_1. Formulas for computing these losses are given in Eqs. (31)–(37) of Chapter 1 by Kreuzer and in the recent article by Kamm (1976). Using Kamm's formulation, the energy dissipated per cycle in a cylindrical container E_L divided by the total acoustic energy of the standing wave E_T takes the simple form

$$E_L/E_T = (2\pi/L)[d_v + (\gamma - 1)\,d_h(1 + L/R)], \tag{23}$$

where L and R are the chamber length and radius, respectively, and the length parameters d_v and d_h are defined by

$$d_v = \text{“viscous boundary layer thickness”} = (2\mu/\rho\omega)^{1/2}, \tag{24}$$

$$d_h = \text{“thermal boundary layer thickness”} = (2k/\rho C_p\omega)^{1/2}, \tag{25}$$

where μ is the gas viscosity. The parameter d_h is simply the inverse of the thermal diffusion parameter α^* defined by Eq. (19); it is the characteristic distance over which the temperature distribution in the gas changes from the constant surface temperature to the oscillating temperature imposed by the acoustic wave. The parameter d_v is defined in an analogous manner with respect to the velocity oscillations induced in the gas. The ratio $(d_v/d_h)^2$ is equal to the Prandtl number $(C_p\mu/k)$, which is 0.7 for air and close to unity for most gases.

Equation (23) provides an explicit form for calculating the decrement of energy per acoustic cycle. The total energy stored in the acoustic mode at any time t', $\Pi_1(t')$, is proportional to the square of the pressure amplitude $A_1(\omega_1;t')$; hence the acoustic decay in the absence of further excitation is

$$\Pi_1(t')/\Pi_1(0) = A_1{}^2(\omega_1;t')/A_1{}^2(\omega_1;0) = \exp[-(E_L/E_T)(\omega_1/2\pi)t']. \tag{26}$$

If only viscous and thermal losses were present, then the logarithmic decrement of Π_1 per cycle would be equal to $Q_1{}^*$:

$$Q_1{}^* = \frac{2\pi}{(E_L/E_T)}. \tag{27}$$

Other losses, particularly those associated with wave scattering of surfaces, also contribute to the signal decrement, and $Q_1{}^*$ from Eq. (27) represents an upper bound to the resonant enhancement that can be achieved. Kamm (1976) has measured Q_1 to be within a factor of two of $Q_1{}^*$ under typical operating conditions.

Equation (26) provides a convenient method of measuring the actual value Q_1 for a spectrophone. The cell is excited at ω_1 by a pulsed excitation source and the value of Q_1 determined from

$$A_1(\omega_1;t')/A_1(\omega_1;0) = \exp(-\omega_1 t'/2Q_1). \tag{28}$$

Equation (28) also describes the behavior of resonant spectrophones with pulsed laser sources provided Q_1 and the coupling between the incident laser pulse and the resonant mode are obtained from calibration experiments.

C. Spectrophones with Wavelength-Modulated Optical Sources

Wavelength modulation is an important and successful technique in optoacoustic spectroscopy. In this section, a simple theoretical model is developed that reproduces the essential features of operation in this mode. It will be found that wavelength modulation is effective in reducing background noise arising from window heating and in discriminating against overlapping absorption lines from interfering species.

Several analyses of derivative spectroscopy have been published (Olson and Alway, 1960; Williams and Hager, 1970; Hager and Anderson, 1970; Penner *et al.*, 1973; and Sulzmann *et al.*, 1973). It is possible (Dewey and Hocker, 1976) to derive a very simple but general model for the interaction between a wavelength-tunable optical source and an absorbing gas by making two assumptions: first, the optical source is narrow with respect to the width of the gaseous absorption line, as is the case with many laser sources; and second, the absorption line has a simple analytic shape. For purposes of illustration, we shall assume a Lorentzian line profile of the form ($\nu \equiv \lambda^{-1}$, cm^{-1})

$$\sigma(\nu) = \sigma_0[1 + (\bar{\nu}/\gamma)^2]^{-1} \tag{29}$$

where σ is the absorption coefficient ($\mathrm{cm}^{-1}\ \mathrm{atm}^{-1}$), σ_0 is the absorption coefficient at the line center ν_0, $\bar{\nu} \equiv (\nu - \nu_0)$ is the difference between the incident radiation wavelength ν (cm^{-1}) and the line center ν_0, and γ (cm^{-1}) is the half-width of the absorption line. Figure 9, from Ku *et al.* (1975), suggests that the Lorentzian model is a reasonable approximation for collision-broadened lines except very far from the line center.

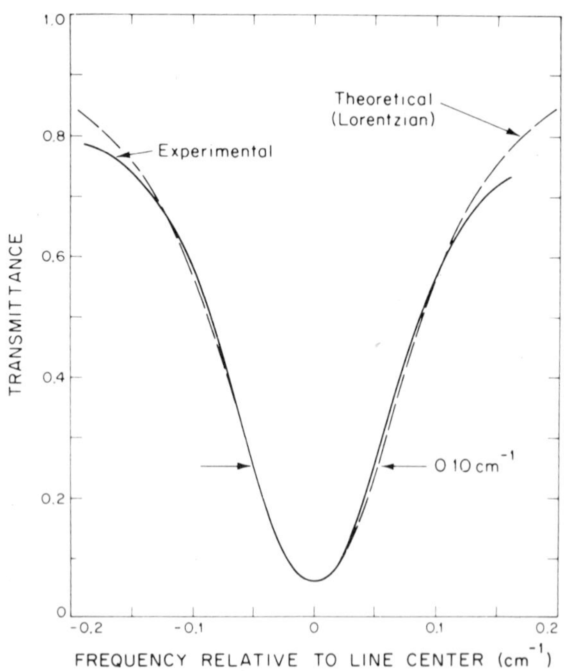

Fig. 9. Measurements of absorption profile of CO P(5) line with a diode laser. (From Ku *et al.*, 1975.)

If the incident radiation is constant in intensity and modulated sinusoidally in wavelength according to

$$\nu = \nu_a + \epsilon \sin \omega t \tag{30}$$

then the time variation of the absorption coefficient is

$$\sigma(t)/\sigma_* = [1 + \alpha_1 \sin \omega t + \alpha_2 \sin^2 \omega t]^{-1}, \tag{31}$$

where

$$\alpha_1 = 2bc/(1 + c^2), \qquad \alpha_2 = b^2/(1 + c^2), \qquad \sigma_* = \sigma_0/(1 + c^2); \qquad b = \epsilon/\gamma, \qquad c = (\nu_a - \nu_0)/\gamma. \tag{32}$$

The term b is a dimensionless quantity, equal to the magnitude of the wavelength modulation divided by the absorption linewidth; and c is the displacement of the central wavelength of the incident radiation from the center of the absorption line, again normalized in terms of the absorption linewidth.

A preferred operating condition is when the incident radiation is modulated in wavelength about the absorption line center. Then $c = 0$, and

$$\sigma/\sigma_* = \sigma/\sigma_0 = [1 + b^2 \sin^2 \omega t]^{-1}. \tag{33}$$

The absorption coefficient σ, which is proportional to the power absorbed by the test gas and hence to the acoustic signal received by the microphone, varies with time and the parameter b^2, as illustrated in Fig. 10. By inspection of the figure, it is clear that, for $c = 0$, there is a substantial component of the microphone signal at twice the modulation frequency f. As c increases, the dominant microphone signal component changes from $2f$ to f (see Fig. 11).

By techniques of Fourier analysis, it is possible to calculate the harmonic content of the microphone signal; thus, for example, for $c = 0$ and $b = 1$ the ratios of successive terms of the Fourier analysis (representing the relative

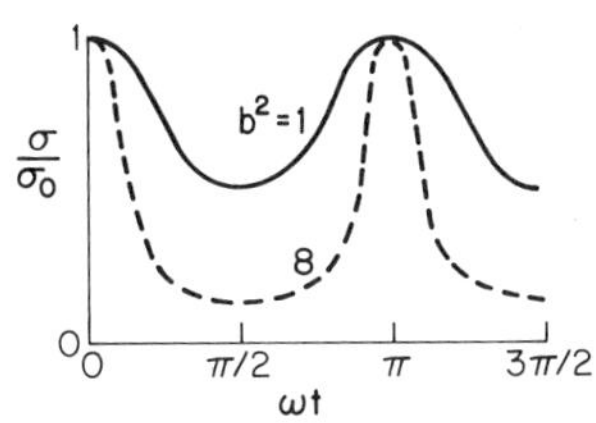

Fig. 10. Modulation of absorption cross section when the wavelength modulation is centered on the absorption line center ($c = 0$).

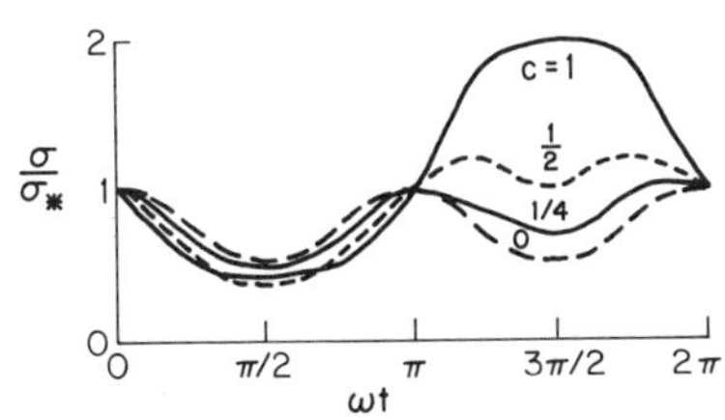

Fig. 11. Variation of harmonic content in the absorption signal as the central wavelength of the wavelength-modulated source is displaced from the center of the absorption line.

magnitudes of signals received by the microphone)* are ($f \equiv \omega/2\pi$):

steady component	:	fundamental at f	:	2nd harmonic at $2f$	:	3rd harmonic at $3f$	:	4th harmonic at $4f$
1	:	0	:	0.242	:	0	:	0.057

It is a general result that the intensity of the microphone signal at f and all odd harmonics ($3f$, $5f$, ...) are zero for $c = 0$. Window absorption and scattering by aerosols, for which $\sigma(\nu) =$ constant, will produce no fluctuating signal.

The common mode of operation with wavelength-tunable sources is to restrict the modulation parameter b to very small values. This limit, derivative spectroscopy, can be recovered from Eq. (31) by taking $b \ll 1$:

$$\frac{\sigma}{\sigma_*} = \left[1 + \left(\frac{2bc}{1 + c^2}\right)\sin \omega t\right]^{-1} \approx 1 - \left(\frac{2bc}{1 + c^2}\right)\sin \omega t. \tag{34}$$

Derivative spectroscopy produces a signal with the major contribution to the fluctuating microphone signal at the modulation frequency f. This signal decreases as c^{-1} for large c, is zero at the center of the absorption line, and shows a maximum at $c = 1$. Long-path absorption measurements of atmospheric pollutants using a tunable diode laser and derivative spectroscopy have been made by Hinkley and Kelley (1971) and Ku *et al.*, (1975); and Patel (1973) has demonstrated its utility with spectrophones.

Although derivative spectroscopy is successful in discriminating against sources of interference having constant absorption coefficients, wide modulation spectroscopy, as described by Eq. (33), is more successful in discriminating against nearby absorption lines of interfering species. A situation encountered frequently in measuring atmospheric pollutants (Gelbwachs, 1974) is that one of the normal atmospheric constituents, "B," such as H_2O, CO_2, or CH_4 will have a strong absorption line near the measurement line of the species of interest, "A." The center of the absorption line of the interferor B is typically displaced from ν_a such that $c_B \gg 1$, and the absorption coefficient modulation in time for species B is

$$\sigma_B(t)/\sigma_B{}^* \approx (1 + \beta \sin \omega t)^{-2}, \tag{35}$$

where $\beta = \epsilon/(\nu_a - \nu_{B0})$ and β is small. A Fourier analysis of Eq. (35) for $\beta \ll 1$ gives

$$\sigma_B/\sigma_B{}^* = 1 - 2\beta \sin \omega t - \tfrac{3}{2}\beta^2 \cos 2\omega t - \beta^3 \sin 3\omega t + \cdots. \tag{36}$$

* The steady component of the absorption coefficient produces heating of the gas which is balanced by steady-state conduction; an ac-coupled microphone will respond only to the Fourier components at f, $2f$, ...

Equation (36) shows that the interferor B produces microphone signals in the ratios:

Steady component	:	Fundamental at f	:	2nd harmonic at $2f$	:	3rd harmonic at $3f$
1	:	2β	:	$\frac{3}{2}\beta^2$	:	β^3

Typical values of β encountered in practice range from 0.1 to 0.01. It is seen, therefore, that the interferor produces a much larger signal at frequency f than at frequency $2f$. By employing wide modulation spectroscopy, Eq. (33), microphone signals are detected at $2f$ and the unwanted signal produced by species B is minimized with respect to the desired signal arising from species A. Conventional derivative spectroscopy employs detection at the fundamental at which frequency the signal produced by the interferor is $(4/3\beta)$ times as large as it is at $2f$.

A detailed examination of Eq. (31) demonstrates that the rejection of unwanted signals from species B depends strongly upon the detailed shapes (Chappell *et al.*, 1971) of the absorbing lines of species A and species B. In certain instances, it may prove desirable to employ detection schemes in which the desired signal is at a frequency f whereas the interfering species contain substantial harmonic content only at higher frequencies. A case in point is detection of the toxic compound nickel carbonyl. This heavy molecule has a very broad absorption line relative to the very narrow absorption lines of interfering species. By employing very large wavelength modulation of the beam, such that ϵ is large compared to the linewidths of the interfering species, the dominant signal contributions of the interfering species will appear at much higher frequencies and detection of the desired signal may be accomplished at either f or $2f$, depending upon the value of c and the nature of the interferors.

Wide modulation spectroscopy provides additional information in the magnitudes of the signals produced at f, $2f$, $3f$, ... which may be used to correct for background signals arising from the interfering species B. As an example, consider the case where the wavelength modulation is centered at ν_0, that is, $c = 0$. Then the signal at $2f$ contains two components, one proportional to $0.242\sigma_0$ and the other proportional to $\frac{3}{2}\beta^2\sigma_{B*}$. At f and $3f$ only signals from the interferor B are present. Therefore, for $\beta \ll 1$, from Eq. (36), the magnitude ratio

$$S \equiv \text{signal at } 3f/\text{signal at } f = \beta^2/2 \tag{37}$$

and the magnitude of β may be determined from the measured ratio S as

$$\beta = (2S)^{1/2}. \tag{38}$$

Therefore, one may calculate the unwanted signal component at $2f$ introduced by species B to be proportional to $3S\sigma_{B*}$ and the signal at $2f$ arising from response to species A may be calculated by

$$T \equiv \frac{\text{signal at } 2f \text{ from A}}{\text{total signal at } 2f} = \frac{0.242\sigma_0}{0.242\sigma_0 - 3S\sigma_{B^*}}, \tag{39}$$

where the minus sign in the denominator indicates that the contributions from species A and B are out of phase. The magnitude of the absorption coefficient σ_{B^*} may be determined in terms of σ_0 (which is known from calibration) by utilizing the measured ratio of signals at f and $2f$:

$$U \equiv \frac{\text{signal at } 2f}{\text{signal at } f} = \frac{0.242\sigma_0 - 3S\sigma_{B^*}}{-2\sqrt{2S}\sigma_{B^*}} \tag{40}$$

or

$$\frac{\sigma_0}{\sigma_{B^*}} = \frac{2(2S)^{1/2}}{0.242}\left[-U + \frac{3}{2\sqrt{2}}S^{1/2}\right]. \tag{41}$$

(Again, the minus signs in Eqs. (40) and (41) are used for out-of-phase terms.) The desired signal ratio T is then

$$T = 1 - (3/U)(S/8)^{1/2} \tag{42}$$

Therefore, the information contained in the measured signals at f and $2f$ may be used to determine the amount of signal at $2f$ which is present because of an interfering species B. This allows the desired signal at $2f$ arising from species A to be computed with substantial accuracy.

Wide amplitude modulation has additional advantages. A calibration cell in series optically with the sample cell and containing a known concentration of species A may be used to lock the center of the modulation wavelength to the line center ν_0 (Dewey and Hocker, 1976). The method may also usefully employ both wavelength and amplitude modulation.

IV. Comparison with Other Analytic Techniques

Optoacoustic spectroscopy is particularly useful with laser optical sources. In conventional absorption spectroscopy, the difference between the incident radiation intensity and transmitted intensity is measured. This difference, as a fraction of the incident radiation, does not increase with the intensity of the excitation source. Fluorescence signals and the pressure fluctuations of optoacoustic spectroscopy increase linearly with the intensity of the exciting source, and these techniques become more sensitive as more powerful optical sources are employed. Attenuation coefficients of 10^{-8} cm^{-1} have been measured in 20-cm-long spectrophones; such sensitivity is impossible with conventional absorption techniques.

In the linear regime, $\sigma L \ll 1$, the optoacoustic signal is directly proportional to the absorbance of the gas. Figure 12, from Patel (1973), is the

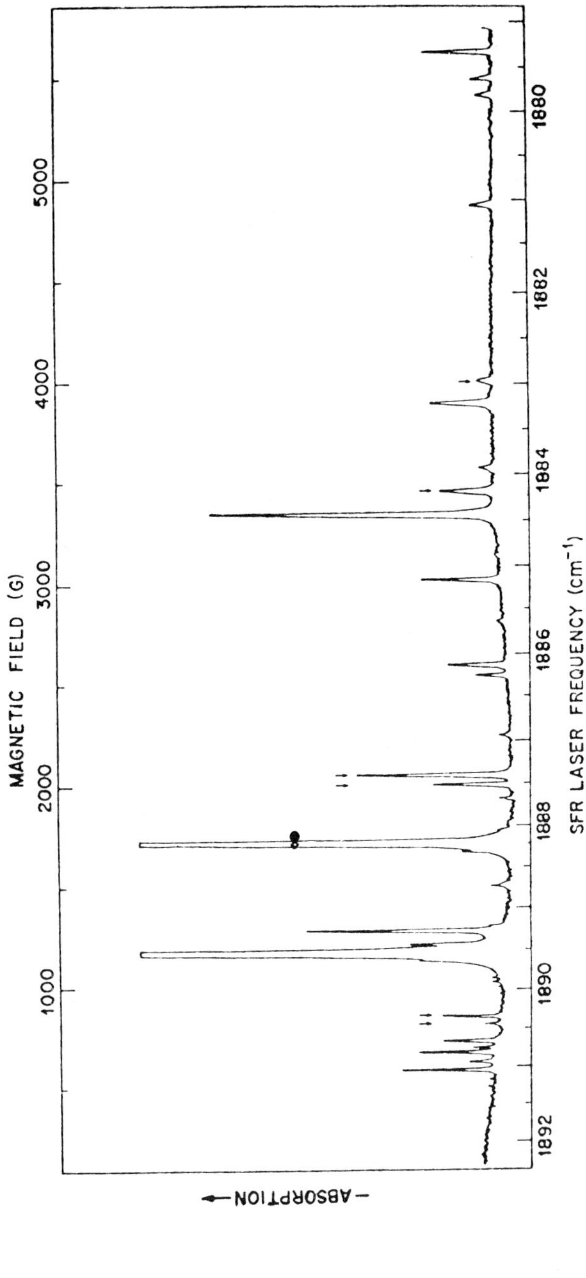

Fig. 12. Optoacoustic absorption spectrum of 20 ppm NO, 76 torr N_2, obtained with a magnetically tuned spin-flip Raman laser; pump laser at 1892.25 cm^{-1}. (From Patel, 1973.)

optoacoustic absorbance spectrum of 20 ppm NO in 76 torr of dry nitrogen obtained with an InSb spin flip Raman laser pumped with a CW CO laser. Arrows indicate NO lines, and the remaining features are from residual water vapor. Figure 13, from Hinkley (1972), compares the conventional absorption spectrum of ethylene (C_2H_4), plotted as percentage of transmission, and the equivalent spectrophone output; these data were obtained using a $Pb_{0.88}Sn_{0.12}Te$ diode laser operating near 10 μm.

The high sensitivity of optoacoustic spectroscopy, when combined with the high resolution available from state-of-the art tunable lasers, offers many advantages for studying the detailed shapes of collision-broadened spectral lines (Dewey, 1974). A very large linear dynamic range, on the order of 10^3–10^5, depending upon the laser source intensity and sample pressure, allows measurements far from the line center.

A frequent application of optoacoustic spectroscopy has been the detection of atmospheric pollutants (see Claspy, Chapter 6, and Kreuter (1974) for accounts of published work). Although wavelength-tunable sources offer many advantages in improving useful spectrophone sensitivity, they are more expensive than fixed-wavelength lasers. Figures 14 and 15 illustrate

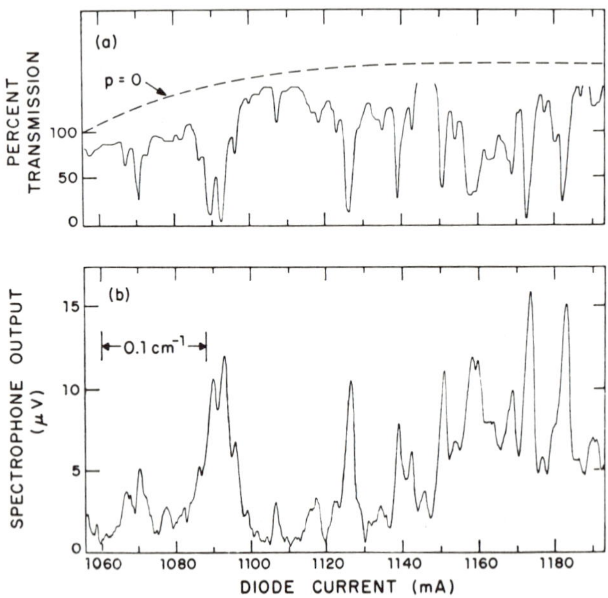

Fig. 13. Comparison between (a) conventional absorption spectroscopy and (b) optoacoustic spectroscopy of ethylene obtained with a semiconductor diode laser. (From Hinkley, 1972.)

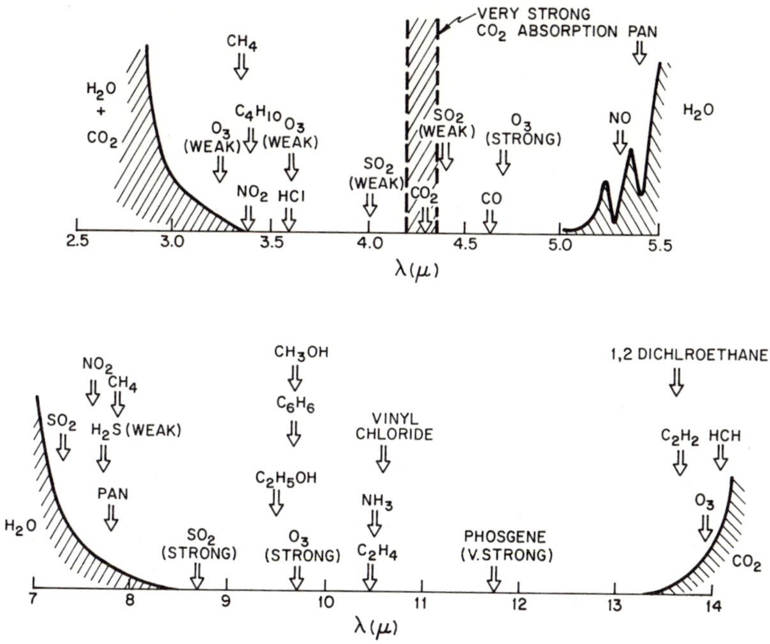

Fig. 14. Diagrammatic representation of the absorption lines of common atmospheric pollutants in the 3–5-μm and 8–14-μm atmospheric transmission windows.

that many common pollutants have useful absorption bands in regions of the infrared spectrum where powerful fixed-wavelength lasers are available.

Gas laser linewidths are generally narrow compared with gaseous absorption lines, and it is therefore necessary to determine with accuracy the relative position of the laser line with respect to the absorption line center and linewidth. The peak absorption near a line center (at least of light molecules) is not decreased by decreasing the total pressure in the spectrophone cell. Since the integral of the absorption cross section over the line is constant, decreasing the linewidth increases the absorption cross section per molecule at the line center. The width of a collision-broadened spectral line is closely proportional to pressure, so that the product of peak absorption cross section and molecular number density remains constant as the pressure of the sample is reduced. Figure 16, from Laguna and Storm (1975), illustrates this behavior for the absorption of 2.744 μm radiation from a HF laser by HF gas. Below about 10-torr pressure, the line broadening is dominated by Doppler effects rather than collisional effects, and the gas absorption coefficient becomes linearly proportional to pressure.

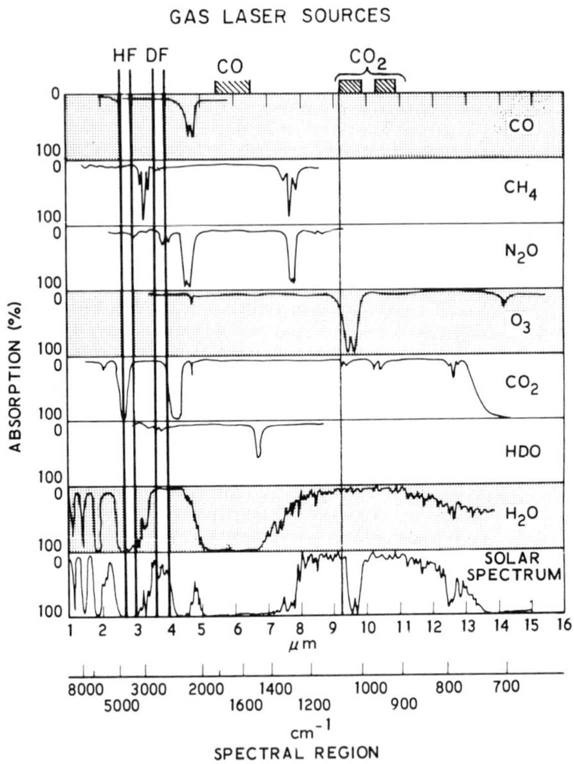

Fig. 15. Infrared spectra of common atmospheric constituents and powerful gas laser sources.

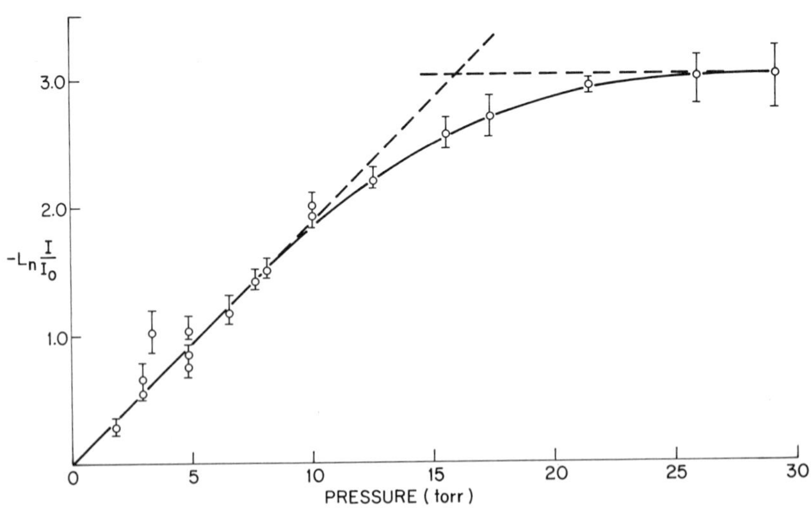

Fig. 16. Absorption of HF laser by HF gas. Below 15 torr, Doppler broadening predominates and the absorption coefficient is proportional to pressure. Above 25 torr, collision broadening yields an absorption coefficient independent of pressure. (From Laguna and Storm, 1975.)

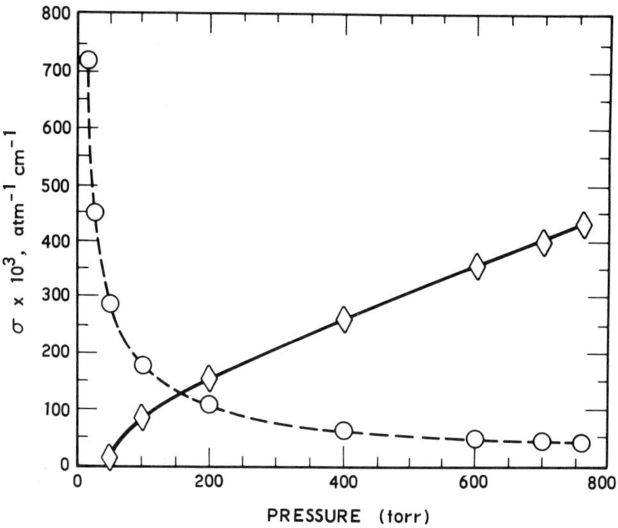

Fig. 17. Variation of absorption coefficient with collisional broadening when the excitation line is nearly coincident with the absorption line center (dashed curve: DF laser line P_1 (9) in CH_4 for $0 \gtrsim P < 1$ atm) and displaced from the line center (solid curve: DF laser line P_2 (11) in N_2O for $0 \gtrsim P < 1$ atm). (From Spencer *et al.*, 1974.)

If the laser wavelength is not coincident with the absorption line center, substantial changes in absorption coefficient can be expected. Figure 17, from Spencer *et al.* (1974), illustrates this behavior. Near the line center, the absorption coefficient per molecule is inversely proportional to pressure, as in the dashed curve. Away from the line center, the absorption coefficient is very low when collision effects are not strong enough to broaden the absorption line to encompass the laser line; this is the behavior illustrated by the solid line.

In conclusion it is useful to point out that blackbody sources may also be useful in optoacoustic spectroscopy. Although several orders of magnitude smaller in power than laser sources, they may be useful in spectrophones that compete favorably with NDIR analyzers at the parts per million level (Dewey, 1974).

Acknowledgments

This research was supported by the Defense Advanced Research Projects Agency (monitored by the Army Research Office, Durham, and the Office of Naval Research) and by the Energy Research and Development Administration.

I am indebted to my colleagues R. D. Kamm and L. O. Hocker for many fruitful discussions. The material on wide modulation spectroscopy, reported here for the first time, was developed collaboratively with L. O. Hocker.

This chapter is dedicated to Robert and Modie Low who contributed their insight and encouragement to scientists throughout the world. All who were privileged to know them mourn their passing.

References

Bates, R. D. Jr., Flynn, G. W., and Knudtson, J. T. (1970). *J. Chem. Phys.* **53**, 3621–3631.

Bauer, H-J. (1972). *J. Chem. Phys.* **57**, 3130–3145.

Bonczyk, P. A., and Ultee, C. J. (1972). *Opt. Commun.* **6**, 196–198.

Bruce, C. W. (1976). Absorption of 9 μm CO_2 laser radiation by atmospheric ozone (unpublished).

Bruce, C. W., Sojka, B. Z., Hurd, B. G., Watkins, W. R., White, K. O., and Derzko, Z. (1976). *Appl. Opt.* **15**, 2970–2972.

Chakerian, C. Jr., and Weisbach, M. F. (1973). *J. Opt. Soc. Am.* **63**, 342–345.

Chappell, W. R., Cooper, J., Smith, E. W., and Dillon, T. (1971). *J. Stat. Phys.* **3**, 401–410.

Claspy, P. C., Pao, Y-H., Kwong, S., and Nodov, E. (1976). *Appl. Opt.* **15**, 1506–1509.

Compton, D. L. (1969). Ph.D. Thesis, Stanford Univ. (unpublished).

Deaton, T. F., Depatie, D. A., and Walker, T. W. (1975). *Appl. Phys. Lett.* **26**, 300–303.

Delany, M. E. (1959). *Sci. Prog.* (*New Haven*) **47**, 459–467.

Dewey, C. F., Jr. (1974). *Opt. Eng.* **13**, 483–488.

Dewey, C. F., Jr., and Hocker, L. O. (1976) (to be published).

Dewey, C. F., Jr., Kamm, R. D., and Hackett, C. E. (1973). *Appl. Phys. Lett.* **23**, 633–635.

Evans, L. B., Bass, H. E., and Sutherland, L. C. (1972). *J. Acoust. Soc. Am.* **51** (Part 2), 1565–1575.

Freund, S. M., and Sweger, D. M. (1975). *Anal. Chem.* **47**, 930–932.

Gelbwachs, J. (1974). *Appl. Opt.* **13**, 1005–1006.

Goldan, P. D., and Goto, K. (1974). *J. Appl. Phys.* **45**, 4350–4355.

Grabiner, F. R., Siebert, D. R., and Flynn, G. W. (1972). *Chem. Phys. Lett.* **17**, 189–194.

Hager, R. N. Jr., and Anderson, R. C. (1970). *J. Opt. Soc. Am.* **9**, 1444–1449.

Hill, D. W., and Powell, T. (1968). "Non-Dispersive Infrared Gas Analysis." Plenum Press, New York.

Hinkley, E. D. (1972). *Opto-electronics* **4**, 69–86.

Hinkley, E. D., and Kelley, P. L. (1971). *Science* **171**, 625–639.

Kaiser, R. (1959). *Can. J. Phys.* **47**, 1499–1513.

Kaldor, A., Olson, W. B., and Maki, A. G. (1972). *Science* **176**, 508–510.

Kamm, R. D. (1973). An acoustic amplifier for the detection of atmospheric pollutants, M. S. Thesis, M.I.T. (unpublished).

Kamm, R. D. (1976). *J. Appl. Phys.* **47**, 3550–3558.

Kamm, R. D., and Dewey, C. F. Jr. (1973). *Proc. IEEE/OSA Conf. Laser Eng. Appl., May 30–June 1*, pp. 35–36.

Kerr, E. L. (1973). *Appl. Opt.* **12**, 2520–2527.

Kerr, E. L., and Atwood, J. G. (1968). *Appl. Opt.* **7**, 915–921.

Kreuzer, L. B. (1971). *J. Appl. Phys.* **42**, 2934–2943.

Kreuzer, L. B. (1974). *Anal Chem.* **46**, 235A, 244A.

Kreuzer, L. B., and Patel, C. K. N. (1971). *Science* **173**, 45–47.

Kreuzer, L. B., Kenyon, N. D., and Patel, C. K. N. (1972). *Science* **177**, 347–349.

Ku, R. T., Hinkley, E. O., and Sample, J. O. (1975). *Appl. Opt.* **14**, 854–861.

Laguna, G. A., and Storm, E. (1975). *J. Appl. Phys.* **46**, 5049–5050.

Max, E., and Rosengren, L-G. (1974). *Opt. Commun.* **11**, 422–426.

Morse, P. M., and Ingard, K. V. (1968). "Theoretical Acoustics," p. 209. McGraw-Hill, New York.

Olson, E. C., and Alway, C. D., (1960). *Anal. Chem.* **32**, 370–373.
Parker, J. G. (1973). *Appl. Opt.* **12**, 2974–2977.
Patel, C. K. N. (1973). *In* "Coherence and Quantum Optics" (L. Mandel and E. Wolf., eds.), pp. 567–593. Plenum Press, New York.
Patel, C. K. N., Burkhardt, E. G., and Lambert, C. A. (1974). *Science* **184**, 1173–1176.
Penner, S. S., Sulzmann, K. G. P., and Chen, H. K. (1973). *J. Quant. Spectrosc. Radiat. Transfer* **13**, 705–707.
Read, A. W. (1967–1968). *Adv. Mol. Relaxation Processes* **1**, 257–272.
Rosencwaig, A., and Gersho, A. (1975). *Science* **190**, 556–557.
Rosengren, Lars-Göran (1973). *Infrared Phys.* **13**, 109–121.
Rosengren, Lars-Göran (1975). *Appl. Opt.* **14**, 1960–1976.
Saltzman, B. E., and Cuddeback, J. E. (1975). *Anal. Chem.* **47**, IR–15R.
Schlichting, H. (1968). "Boundary-Layer Theory." McGraw-Hill, New York.
Schnell, W., and Fischer, G. (1975a). *ZAMP* **26**, 133–135.
Schnell, W., and Fischer, G. (1975b). *Appl. Opt.* **14**, 2058–2059.
Sulzmann, K. G. P., Lowder, J. E. L., and Penner, S. S. (1973). *Combust. Flame* **20**, 177–191.
Spencer, D. J., Denault, G. C., and Takimoto, H. H. (1974). *Appl. Opt.* **13**, 2855–2868.
Tripodi, R. (1970). *J. Chem. Phys.* **52**, 3298–3299.
Tripodi, R., and Vincenti, W. G. (1971). *J. Chem. Phys.* **55**, 2207–2217.
Williams, D. T., and Hager, R. N., Jr. (1970). *Appl. Opt.* **9**, 1597–1605.
Woodmansee, W. E., and Decius, J. C. (1972). *J. Chem. Phys.* **36**, 1831–1837.

4

Tunable Radiation Sources in the Ultraviolet and Visible Spectral Regions (0.1–1.0 μm)

Jerry A. Gelbwachs

Chemistry and Physics Laboratory
The Aerospace Corporation
El Segundo, California

I. Introduction

The photoacoustic effect was demonstrated during the latter part of the nineteenth century, but, after the initial excitement subsided, the method rapidly fell into disuse. In the 1970s, the method was revived primarily because of technological advances that had occurred during the preceding two decades. One of these important achievements was the development of the laser. Compared to incoherent sources of radiation, laser sources offer (1) an enormous improvement in collimated optical intensity per unit optical bandwidth, and (2) greater wavelength coverage. This chapter describes tunable sources of radiation in the 0.1–1.0-μm spectral region with emphasis upon laser sources, which are applicable to photoacoustic spectroscopy.

Incoherent emitters of continuum radiation are discussed in Section II. Section III begins with a brief review of the principles of stimulated emission and the characteristics of laser radiation. Four discrete-wavelength lasers are then discussed with respect to the properties that have made them particularly useful excitation sources for the generation of tunable coherent emission. Probably the most significant source of tunable near-ultraviolet, visible, and near-infrared radiation for spectroscopic applications is the dye laser. Its characteristics and tuning properties are discussed at the end of Section III. The more important techniques for extending the wavelength coverage of tunable coherent emission into the ultraviolet region are reviewed in Section IV. These methods are based upon nonlinear optical phenomena and include second-harmonic generation, frequency mixing, optical parametric oscillators, and four-wave mixing in metal vapors and inert gases. The topics discussed in this chapter are summarized in Section V.

Besides the sources mentioned above, other sources of tunable radiation exist in the 0.1–1.0-μm spectral region. Among them are light-emitting diodes, injection lasers, excimer lasers, and stimulated Raman scattering. However, at this writing, each of the alternative sources is inferior to the above on the basis of spectral coverage and/or output power. Major technological advances in the state of the art of these devices are necessary before they become competitive with the methods discussed in this chapter. Therefore, owing to their limited capacity for the generation of tunable light and constraints placed upon the length of this chapter, no further mention to these alternate sources will be made herein.

II. Incoherent Sources of Continuum Radiation

Features of the more common sources of continuum radiation for photoacoustic spectroscopy in the 0.1–1.0-μm spectral region are briefly discussed in this section. A survey of incoherent ultraviolet, visible, and infrared emitters is given by Cann (1969). Koller (1965) has reviewed ultraviolet sources; those that emit at wavelengths shorter than 200 nm have been treated by Samson (1967) and McNesby *et al.* (1971). These sources fall into two major categories: incandescent emitters and arc sources.

The emission from an incandescent source is associated with its temperature and can be approximated by blackbody radiation. The total energy emitted per unit area of a blackbody source is proportional to the fourth power of its temperature according to the Stefan-Boltzmann law (Kasper, 1972). The spectral distribution of blackbody radiation is given by Planck's radiation formula.

The most familiar source with an incandescentlike emission spectrum is the sun and warrants special mention in a book about photoacoustic spectroscopy. The sun was the source in the initial demonstration in 1881 of the photoacoustic effect by Alexander Graham Bell. Its spectral characteristics are well known and have been carefully tabulated (Billings, 1972). On the earth, the sun provides continuum optical radiation that is economical and pollution-free from 300 nm to beyond 1 μm. Its spectral output closely matches that of a 5900°K blackbody. At sea level, the emission peaks near 500 nm. Absorption by atmospheric gases effectively blocks solar radiation at wavelengths shorter than 300 nm.

The tungsten lamp provides a simple, low-cost source of continuum radiation throughout the visible and infrared wavelength regions. The standard coiled-coil tungsten filament is contained in a quartz envelope and operates at a temperature of 3000°K. The addition of a halogen gas to the quartz bulb reduces the rate of tungsten vaporization and prolongs the life of the lamp to beyond 1000 hr. Its spectrum closely corresponds to that of a blackbody at 3000°K. Greater emission is possible by operating the lamp at higher temperatures though lifetime is sacrificed as the melting point of tungsten (3655°K) is approached.

Optical radiation can be obtained from a gas or vapor through which an electrical current is passed. Arcs are discharges through a gas or a vapor that is usually contained in glass or quartz envelopes. The pressure inside the envelope may vary from a few millitorr to hundreds of atmospheres and is limited only by the mechanical strength of the envelope. Low-pressure arcs emit radiation at discrete wavelengths characteristic of the particular gas in the discharge. A continuum emission from the higher-lying states of certain gases is observed from high-pressure arcs. Large luminuous efficiencies have been shown to accompany high-pressure operation. Because high-pressure operation raises the temperature of the bulb, forced air cooling or water cooling is necessary to prevent the lamp from overheating and softening. The emission from arc sources is nonuniform with maximum intensity near the cathode and along the arc axis.

Xenon arc lamps operate at pressures of 50–70 atm and are efficient emitters of intense near ultraviolet, visible, and near infrared continuum radiation. Some very intense lines are generated between 0.9 and 1.2 μm. The discharge is usually enclosed in small spherical or ellipsoidal quartz bulbs and is operated at arc temperatures between 8000 and 10,000°K. Commercial sources range in size from 75 to 30,000 W. The arc stability is greater when the lamps are operated on direct current. The surface brightness of the arc is somewhat lower but more uniform with alternating current excitation.

The high-pressure mercury lamp is an extremely intense source of multi-line radiation at wavelengths between 230 and 650 nm. Mercury vapor lamps come in a wide variety of physical shapes and sizes. These factors have little influence upon the nature of the emitted radiation. The 1000-W mercury arc generates more than 170 W throughout the ultraviolet and 240 W in the visible region. The small diameter (0.1–5 mm) of the arc is useful for imaging in optical systems.

The carbon arc is an open electrical discharge between two electrodes in the ambient air. During operation, the electrodes are consumed and require replacement after several hours. Compared to enclosed arcs, the carbon arc is inherently less efficient although it emits energy from a smaller area. The carbon arc emission consists of the continuous spectrum of the incandescent electrodes corresponding to blackbody temperatures of 4000–6000°K. Superimposed upon the continuum is a strong feature at 388 nm owing to the emission from the CN violet bands. Various carbon mixtures can be used as electrodes to generate radiation with slightly differing spectra (Koller, 1965). Strontium-cored carbon arcs display enhanced emission near 700 nm.

No single source of intense continuum radiation exists in the wavelength region below 200 nm. A variety of sources produce continuous spectra over limited wavelength regions. For the most part, the output power and lifetime of these sources are less than for the lamps previously discussed with the exception of the carbon arc.

The hydrogen and deuterium arcs provide a line-free continuum between 165 and 250 nm. High-pressure water-cooled deuterium lamps are convenient to work with and commercially produced. Rare-gas capillary lamps, (Tanaka, 1955) and microwave electrodeless lamps at 2.45 GHz have proved useful as moderately intense sources below 200 nm. The xenon continuum provides radiation between 150 and 225 nm (Wilkinson and Tanaka, 1955), while the krypton continuum covers 125–165 nm (Wilkinson, 1955). A list of the above incoherent sources of continuum radiation and their useful wavelength regions appears in Table I.

To obtain photoacoustic spectra, isolation of a narrow band of optical radiation from the emission of the continuum source is required. Low-resolution spectra (>10 nm) can be acquired by the use of various types of

TABLE I

Incoherent Sources of Continuum Radiation (0.1–1.0 μm)

Continuum light source	Wavelength region (nm)	$\log_{10}$ Peak spectral radiance [mW $(mm^2 \cdot sr \cdot nm)^{-1}$]
Sun (sea level)	300– > 1000	+1 at 550 nm
Tungsten filament lamp (1 kW)	600– > 1000	0 at 800 nm
Xenon arc lamp (1 kW)	400–2500	+1 at 550 nm
Mercury arc lamp (1 kW)	230–650	+1 at 400 nm
Carbon arc (cerium core, 60 A)	350–700	+1 at 400 nm
Hydrogen lamp (1 kW)	165–250	−1 at 230 nm
Xenon continuum lamp	150–225	−2 at 170 nm
Krypton continuum lamp	125–165	−3 at 150 nm
[Dye laser (20 mW)]	[340–1200]	[+10]

materials that filter light by absorption, reflection, interference, and scattering (Calvert and Pitts, 1965). Optical filters are made from glass, gelatin, thin films, plastics, and dissolved organic salts. Incoherent continuum sources provide satisfactory excitation for the generation of low-resolution spectra in optical systems requiring a minimum of source imaging.

High-resolution (< 1 nm) spectra are routinely achieved by the use of a continuum radiation source in conjunction with a monochrometer. A useful quantity for the comparison of light sources based upon the energy that may be transmitted through a monochrometer is the power radiated per unit area, per solid angle, and per unit optical bandwidth. This measure of source brightness is called the spectral radiance. It is an invariant of an optical system, that is, the brightness of the image may never exceed that of the source (Born and Wolf, 1965). The approximate spectral radiance of the continuum sources previously discussed is listed in Table I along with the value for a 20-mW tunable laser. It can be seen from Table I that the spectral radiance of the 20-mW laser greatly exceeds that of the incoherent continuum sources, which indicates its enormous advantage for tunable, high-resolution spectroscopic work. The exceptional brightness of the laser is a result of the high degree of monochromaticity and spatial coherence that arise during the process of stimulated emission. However, at this writing, the cost of lasers is measured in thousands of dollars, whereas lamps can be purchased for several hundred dollars.

In conclusion, incoherent sources of continuum radiation provide intense emission from 230 to 1000 nm. The lamps that radiate in this spectral region are compact, reliable, rugged, and relatively inexpensive. Emission at wavelengths shorter than 230 nm is less intense and requires several sources to span the 110–230-nm region. Incoherent sources are effective for the acquisition of low resolution spectra of large samples (several square centimeters).

III. Laser Sources

A. Introduction

Its ability to generate optical radiation that is confined to extremely narrow spectral widths and small spatial volumes has made the laser an invaluable tool to scientists for the study of material properties. Since 1960, a vast body of knowledge has been accumulated in the areas of laser physics and quantum electronics. The intent of this section is to provide a working knowledge of laser principles to the experimenter whose expertise lies in nonlaser-related areas so that he can better determine the type of laser that meets his particular requirements. An understanding of these concepts will also enable the experimenter to compare commercial laser systems and to use laser sources effectively. For a complete treatment of the topics briefly discussed here, the reader is encouraged to consult the many excellent texts on lasers and quantum electronics (Lengyel, 1966; Pantell and Puthoff, 1969; Chang, 1969).

B. Laser Concepts

1. Laser Requirements

In its simplest form, the laser, or optical oscillator, consists of an active medium that supplies gain for optical radiation, a method for exciting the active medium, and a feedback mechanism. Gain is provided at optical wavelengths by two states of the active medium (1) that are coupled through an electric dipole matrix element and (2) in which the upper level can be preferentially populated to create a population inversion. The gain process is the exact inverse of optical absorption. Transitions among electronic states, vibrational states, and rotational states have produced stimulated emission in the ultraviolet, visible, infrared, and far-infrared portions of the electromagnetic spectrum. Laser emission has been observed in solids, gases and vapors, and liquids. There are over 350 different discrete laser wavelengths in the 120-nm to 1-μm spectral region (Pressley, 1971).

The common methods of preparing the active medium such that it exhibits gain include electrical discharge, optical pumping by various lamps or other lasers, exothermic chemical reactions, and electron beam excitation.

The active medium is placed inside a resonator formed by reflective mirrors to provide the feedback of radiation for the optical oscillator. To couple energy out of the resonator, one mirror is partially transmitting. The condition at which gain equals losses is called threshold. The primary resonator loss mechanisms are absorption and scattering by optical components in the resonator and output mirror transmission.

2. Spectral Characteristics

The spectral output of the laser can be better understood with the aid of Fig. 1 in which optical gain and loss is plotted versus optical frequency ν. For laser action to occur, the gain for light passing through the active medium must exceed the losses experienced traversing the optical cavity. The cavity losses are approximately independent of optical frequency and are indicated by a horizontal line in Fig. 1. The gain profile of the active media is usually a bell-shaped curve that can be mathematically described by a Gaussian profile or a Lorentzian distribution (Yariv, 1967). Its full width at half maximum is given by $\Delta\nu_a$. Oscillation may occur only at frequencies for which the gain exceeds the cavity losses, namely, between ν_1 and ν_2.

For optical fields inside the resonator to experience gain, radiation propagating along the axis of the laser cavity must satisfy the conditions for constructive interference after a round-trip through the cavity. Consequently, the radiation is compelled to exist only at discrete frequencies called cavity modes. For a cavity of optical length L, the frequency spacing between adjacent cavity modes is $c/2L$, where c is the speed of light. For example, the intermode spacing of a 30-cm optical path length cavity is 500 MHz.

The number of laser modes equals the number of cavity modes between ν_1 and ν_2. Certain laser materials, such as those used in low-pressure molecular lasers, have narrow natural linewidths; therefore, only one cavity mode has sufficient gain to reach threshold. The ultimate linewidth of a single mode is determined by quantum effects, whereas the observed linewidth is broader and stems from variations in the optical length of the reasonator. The magnitude of the frequency variations caused by quantum noise has been studied in great detail (Yariv, 1967). The minimum theoretical bandwidth of a single mode is of the order of 1 Hz over a 1-sec time duration. Most lasers have a cavity structure consisting of two end mirrors held together by mechanical supports. The structure is subject to bending and stretching from mechanical

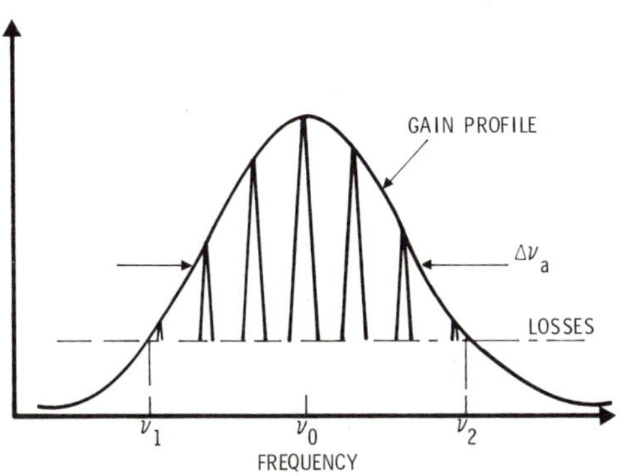

Fig. 1. Optical gain and losses in a laser resonator versus optical frequency.

vibrations, and length variations from thermal changes. These environmental factors alter the cavity length, thereby causing variations in the laser frequency. They have been shown to impose limits that preclude realization of the quantum noise limit. In practice, a spectral purity of 1 part in 10^7 is easily attainable and frequency-stabilized lasers have demonstrated stabilities of 1 part in 10^{14}. Pulsed lasers are more difficult to stabilize than CW (continuous-wave) lasers because of temperature gradients induced by heating of the cavity components.

3. *Mode Locking*

When several cavity modes oscillate simultaneously, the amplitude and phase of the oscillations vary randomly with respect to one another and cause instantaneous variations in the output power of the laser. Methods are available to impose fixed relationships upon the amplitude and the phase of the individual cavity oscillations. This process is known as *mode locking* and produces laser outputs that are well defined functions of time. Mode locking is reviewed by DeMaria *et al.* (1969) and Smith (1970). Mode-locked lasers usually emit energy in the form of high-power spikes periodically spaced in time with a period of $2L/c$, the roundtrip cavity time. The duration of each pulse can be as short as the reciprocal of the natural linewidth, and pulse widths of 10^{-9}–10^{-13} sec have been recorded; the shortest pulses have been derived from glass and dye lasers. Peak powers of 10^9 W are routinely produced by this technique.

Several mode-locking methods are commonly employed to generate subnanosecond pulses. In the first, a phase modulator consisting of an electrooptic crystal is placed near one cavity mirror. Application of a sinusoidal electric field at the intermode frequency creates a time varying optical phase shift. Groups of photons that pass through the electrooptic crystal experience phases shifts dependent upon the value of instantaneous applied voltage. After successive traverses through the modulator, the apparent frequency shift will be sufficient to place the photons outside the gain curve resulting in no further gain. Only those photons that pass through the crystal at times when the voltage is zero continue to experience amplification. Consequently, the output is composed of a series of pulses separated by $2L/c$. This method of mode locking is termed *phase modulation.*

The second method uses an acoustooptic crystal. It resembles the first in that the crystal is placed at one end of the cavity and is excited by an acoustic signal at the intermode frequency. Bragg diffraction of the optical radiation by the acoustic waves introduces loss to light passing through the crystal at all times except when the driving voltage passes through zero. The laser will seek to operate under the conditions of least loss, which corresponds to an output consisting of pulses spaced by $2L/c$. Locking the modes of a laser in this manner is known as *loss modulation.*

A saturable absorber located inside the laser cavity is also an effective mode locker. A saturable absorber is a material, usually a dye in liquid solution, for which the absorption constant decreases with increasing light intensity caused by saturation. The condition of minimum loss favors an output in which the energy is emitted in the form of periodic high-power pulses rather than a continuous wave. Again a repetitive train of pulses is obtained at the $c/2L$ rate. This third method of mode locking is passive and does not require external control.

4. Q Switching

High-energy pulses are obtained by Q switching (Hellwarth, 1966), a technique that involves switching the cavity resonator from a high-loss to a low-loss configuration. Laser oscillation commences when the population inversion (or gain) exceeds the threshold value (determined by the cavity losses). If the cavity losses are temporarily increased, or the cavity Q is spoiled, oscillation is prevented. When the inversion reaches a peak, the cavity Q is suddenly restored to its former high value and the laser gain is now well above threshold. A rapid depletion of the population inversion ensues that leads to a buildup of radiation inside the laser cavity that appears at the output as a single giant pulse of short duration. The available output power is determined by the energy stored in the population inversion. The method is effective for laser materials in which the upper state of the laser transition is long-lived, that is, on the order of milliseconds. Q-switched ruby and Nd:YAG solid-state lasers produce 1–10-J pulses in 10–100 nsec. Rotating mirrors, electrooptic shutters (Kerr cells and Pocket cells), acoustooptic devices, and saturable absorbers are common Q-switching elements.

The finite buildup time of the field inside the laser cavity and the time required to repump the inversion set an upper limit upon the repetition rate of a Q-switched laser. The maximum rate is of the order of the reciprocal of the radiative lifetime of the upper state.

5. Cavity Dumping

Cavity dumping of a continuously pumped laser is a method to enable extraction of pulses of energy from a laser cavity at higher repetition rates than those achieved by Q switching. An acoustooptic deflector is placed at the Bragg angle inside a totally reflecting laser cavity. Pulses are periodically introduced to the modulator launching acoustic waves transverse to the direction of beam propagating. The acoustic waves deflect light from the cavity by Bragg diffraction. As much as 50% of the ultracavity energy can be extracted by this method.

Oscillations are never totally quenched as in the Q-switched mode of operation. Consequently, subsequent pulses can be delivered in durations shorter than the time required for the field inside the cavity to build itself from the noise level. However, as more energy is extracted per pulse, the buildup time of the energy inside the cavity does increase, thus an inverse relationship exists between maximum repetition rate and peak power output. The repetition rate of light pulses from a cavity-dumped Nd: YAG laser has been controlled from 125 kHz to several megahertz with a minimum-pulse duration of 30 nsec (Maydan and Chesler, 1971). A continuous wave (CW) 1-W argon ion laser has generated 0.5-nsec pulses with peak powers of 60 W at a frequency of 2 MHz.

A single intracavity acoustooptic modulator may provide a laser output that is both mode locked and cavity dumped.

6. *Spatial Properties of a Laser Beam*

The laser cavity controls the excellent spatial coherence of the laser, which determines how tightly the beam can be focused and how well it can be collimated. The spatial distribution of electromagnetic radiation inside a cavity bounded by reflective surfaces has been extensively studied (Fox and Li, 1960; Boyd and Gordon, 1961; Boyd and Kogelnik, 1962). Besides being quantized in the longitudinal direction as determined by the interference conditions, the optical waves are also quantized in the transverse direction. The resulting modes are designated TEM_{nm} and TEM_{pl} for rectangular and cylindrical geometries, respectively. Modes of circular geometry are not commonly observed in practice unless the resonator is carefully adjusted to be symmetric about the cavity axis.

The most important resonator mode is the fundamental TEM_{00}; its energy is concentrated near the resonator axis and it suffers the smallest diffraction loss upon passage through finite diameter optical components. The beam intensity in the transverse direction exhibits a Gaussian profile. The beam radius w is defined as the distance at which the field amplitude is $1/e$ times that on the axis. In terms of beam intensity, a quantity more amenable to measurement, it is the distance at which the intensity falls to e^{-2} of its maximum value: 87% of the energy is contained within the beam radius. A schematic representation of a Gaussian beam appears in Fig. 2. The beam is completely characterized by the confocal parameter and the distance to the minimum beam waist w_0. The confocal parameter b, defined by $b = 2\pi w_0{}^2/\lambda$, is a measure of the distance over which the beam remains essentially collimated. At a longitudinal distance $z = b/2$ from w_0, the beam area has expanded from its minimum value by a factor of 2. For

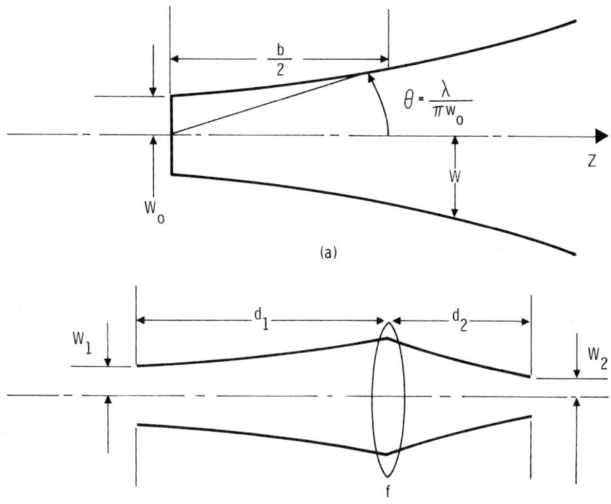

Fig. 2. (a) Parameters of a TEM_{00} Gaussian beam: the beam waist is w; b is the confocal parameter; and θ is the beam divergence. (b) Geometry for transforming a Gaussian beam with waist w_1 to one with waist w_2 by the use of a lens with focal length f.

$z > b/2$, beam divergence is approximately given by the far-field diffraction angle $\theta = \lambda/\pi w_0$. Typical values for a visible laser operating in the TEM_{00} mode are $\lambda = 500$ nm and $w_0 = 1$ mm. The corresponding values for confocal parameter and beam divergence are $b = 12$ m and $\theta = 0.2$ mrad.

For some applications, it is necessary to focus the laser beam to a new waist as shown in Fig. 2b. The minimum focal length lens f_0 that will transform the beam waist is given by $f_0 = (b_1 b_2/4)^{1/2}$, where b_1 and b_2 are the confocal parameters for the original and transformed beam, respectively (Kogelnik and Li, 1966). For a lens whose focal length f exceeds f_0, the distances d_1 and d_2 between the lens and the beam waists are adjusted in accordance with

$$d_1 = f \pm (1/2b_1)[(f/f_0)^2 - 1], \qquad d_2 = f \pm (1/2b_2)[(f/f_0)^2 - 1].$$

C. Discrete-Wavelength Lasers for Pumping Tunable Coherent Sources

Laser emission has been recorded throughout the ultraviolet, visible, and infrared spectral regions. The stimulated output of most lasers occurs between well-defined energy levels of the laser medium generating emission that appears as discrete lines at low spectral resolution. Tuning of discrete wavelength lasers is limited to their natural linewidths: 1 part in 10^6 for gas lasers, 1 part in 10^3 for solid-state lasers, and 1 part in 10^2 for glass systems.

The narrow spectral widths of discrete-wavelength lasers are unsuitable for spectroscopic studies requiring tunable excitation sources.

Four discrete-wavelength lasers thus far have proved useful for exciting broadly tunable sources of coherent visible and ultraviolet radiation. The lasers are the ruby laser, the Nd:YAG laser, the N_2 laser, and the argon ion laser. A complete description of each is beyond the scope of this article. Only the features that pertain to their role as excitation sources for tunable sources will be mentioned. In-depth treatments of these lasers can be found elsewhere (Smith and Sorokin, 1966; Siegman, 1971; Yariv, 1971). The sources of tunable coherent radiation are dye lasers, second-harmonic generators, optical parametric oscillators, frequency mixers, and third-order nonlinear optical generation in metal vapors and inert gases. These devices are described in subsequent sections.

1. Ruby Laser

Ruby was the first material to exhibit laser emission (Maiman, 1960). The terminal laser level is the ground state; thus, intense pumping is needed to establish a population inversion. The laser is relatively inefficient and is usually operated in a pulsed mode. The long radiative lifetime of the upper level (3 msec) allows energy storage in the inversion that enables peak powers of 10^8–10^9 W to be delivered in 10 nsec by Q-switching methods. The laser wavelength is 694 nm at room temperature. Q-switched pulses at the ruby fundamental frequency and second harmonic can pump tunable dye lasers and optical parametric oscillators.

2. Nd:YAG Laser

The solid-state Nd:YAG laser radiates at 1.06 μm. The laser gain is large ($\sim$75 times that of ruby) and, at room temperature, the population of the lower laser level is negligible, which allows threshold to be easily reached, both on a pulsed and a continuous-wave basis. The 230-μsec lifetime of the upper laser level is sufficient for the Q-switched generation of 10^8-W pulses in 10 nsec. The oscillation threshold for the Nd:YAG laser is approximately three orders of magnitude lower than that of ruby. Also, its low threshold and good thermal conductivity allow pulsed operation at higher repetition rates. The second harmonic of Nd:YAG at 532 nm pumps visible dye lasers and parametric oscillators, whereas the third harmonic at 354 nm excites ultraviolet dye lasers. Nd:YAG laser emission can be efficiently converted to radiation at 266 nm by doubling the second harmonic frequency. High peak power pulses at 266 nm are useful for the sum frequency generation between 200 and 250 nm and for phase-matched third-order nonlinear processes that produce tunable coherent emission in the 100–200-nm spectral region.

3. *Nitrogen Laser*

The nitrogen laser operates between levels in the second positive system of N_2 (Heard, 1963; Gerry, 1965). The radiative lifetimes of the upper state C $^3\Pi_u$ and lower state B $^3\Pi_g$ are 40 nsec and ~10 μsec, respectively. The lifetimes are just opposite to those required for optimum lasing and consequently emission only occurs in times shorter than the upper state lifetimes because of the bottleneck effect (population buildup) at the lower laser level. A variety of fast electrical discharge circuits provide rapid excitation of the upper state (Leonard, 1965; Shipman and Kolb, 1966). The N_2 laser has emitted peak power pulses of up to 10 MW in 10 nsec at 337 nm (Godard, 1974). Commercial units deliver 1 MW in 10 nsec at a rate of 100 pulses per second (pps) and are employed for pumping dye lasers.

Because the short radiative buildup time in the optical cavity is insufficient for the development of well-defined cavity modes, the emission at 337 nm is spread continuously over 0.2 nm. The rectangular discharge channel creates a beam with a typical cross section of 10 × 20 mm. This rectangular beam is suitable for side pumping the dye laser in contrast to the TEM_{00} modes of the Nd:YAG, ruby, and argon ion lasers that normally optically excite the dye along the cavity axis.

Recent work has shown that N_2 laser emission can be sustained for times longer than the upper state lifetime. The addition of SF_6 to N_2 produced a 75-nsec pulse and radiation at 358 as well as 337 nm (Suchard *et al.*, 1975).

4. *Argon Ion Laser*

An electrical discharge in gaseous argon creates inversion between many levels of doubly ionized argon, singly ionized argon, and atomic argon. Pulsed and CW stimulated emission has been observed at discrete wavelengths spanning 262–1092 nm (Bridges and Chester, 1971). Highly stable, CW outputs occur at several ultraviolet wavelengths near 360 nm and at a dozen blue-green wavelengths. Commercial units emit 20 W of CW-visible TEM_{00} power that is convertible with up to 30% conversion efficiency into dye laser emission across the 510–700-nm wavelength band. More than 4 W of stable argon ion ultraviolet power is present to pump blue and green CW dye lasers.

D. Dye Lasers

The development of tunable dye lasers has literally revolutionized the field of visible and ultraviolet spectroscopy. Today, by the use of selected dyes, the wavelength of dye lasers can be continuously tuned over the

340–1200-nm region with a spectral width less than 1 nm (Deutsch, 1971). Each dye lases over ~50–100 nm. One part in 10^9 resolution, far superior to that attained by ordinary spectroscopic methods, is possible by the insertion of optical elements inside the dye laser cavity (Wu *et al.*, 1974). In addition to its role as a primary source of continuously tunable narrow band optical radiation, the dye laser can serve as the excitation source for devices that generate tunable coherent emission by nonlinear processes (see Section IV). When the dye laser assumes the latter function, continuously tunable coherent radiation has been generated to wavelengths as short as 100 nm. In this manner, coverage is possible of wavelength regions where strong incoherent sources are difficult or impossible to construct. The tunability, high spectral purity, and superior spatial coherence of dye lasers provide an excellent source for high-resolution spectroscopy and eliminate the need for bulky conventional spectroscopic apparatus that includes high-intensity lamps, lens systems, monochromators, and spectrometers. However, dye lasers tend to be expensive, and the lifetime of some dyes is as short as several hours.

The dye laser combines many advantages of solid-state and gas lasers. Its gain may exceed that of many solid-state laser materials by several orders of magnitude. The use of a liquid active medium guarantees high optical quality, facilitates cooling, resists damage from intense optical fields, and is less costly than crystalline laser materials.

1. Energy Levels of Dyes

Figure 3 is an energy level diagram of a representative laser dye molecule in solution. Two manifolds of electronic levels are the singlet and triplet states designated by S and T. Transitions between S and T states are spin forbidden. Quantum mechanics has shown that, for every excited singlet state, there exists a triplet state of somewhat lower energy. For large molecules such as dyes, many normal modes of vibration of differing frequencies are coupled to the electronic transitions. The vibrational energies are of the order 1000 cm^{-1}. Superimposed upon the vibrational levels are a series of rotationally excited sublevels with splittings of the order of 10 cm^{-1}.

The widths of the rotational levels are extremely broadened by collisions with solvent molecules, which produces a quasi-continuum of levels associated with every electronic level. Collisions with solvent molecules that occur on the order of 10^{-11}–10^{-12} sec at room temperature also produce rapid thermalization within individual electronic levels. Thus, the absorption (depicted by the upward arrow from S_0 to S_1) and fluorescence are practically continuous over the absorption band.

Upon excitation to a high-lying vibrational–rotational state in $S_1$④, the molecule rapidly relaxes by nonradiative processes to the lowest level of $S_1$③.

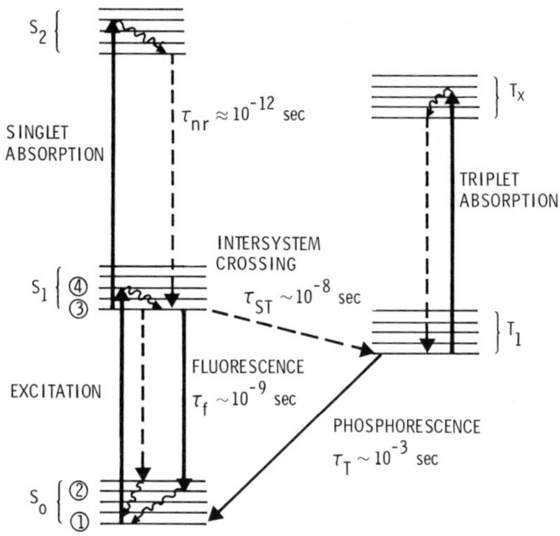

Fig. 3. Energy level diagram of a dye molecule in solution. Singlet levels and triplet levels are designated S and T, respectively.

The dye molecule may decay from ③ by fluorescence emission to any of the sublevels in the ground state manifold, thus giving rise to a continuum fluorescence spectrum. The emission band is a mirror image of the first excited electronic state absorption band through the wavelength that corresponds to the transition between the lowest in S_0 and the lowest state in S_1(① → ③) (Fig. 4). The separation between the absorption peak and the emission peak is ~ 50 nm for typical dyes. The emission bandwidth generally varies between 50 and 100 nm.

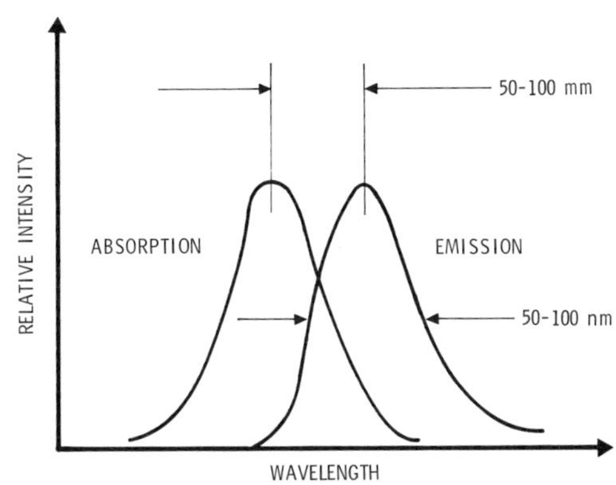

Fig. 4. Absorption and emission spectrum of a dye molecule. The intensity scale is different for each spectrum.

Fluorescence from ③ to S_0 occurs within several nanoseconds, compared with milliseconds for solid-state laser materials. The short fluorescence lifetime τ_f of dyes makes it difficult to store energy in the population inversion. Therefore dyes are poor candidates for *Q*-switched lasers. Laser action between States ③ and ② of Fig. 3 requires population inversion between the states. If population inversion is to be attained (1) state ③ must be pumped fast enough so that its population exceeds the thermal population of ② and (2) ② decays to ① faster than ③ decays by all processes. The first requirement defines lasing threshold. The second is a function of the energy levels and decay channels of the dye molecule in solution. As discussed, the rapid nonradiative decay of an excited state within an electronic manifold aids the accumulation of population in ③ and the removal of molecules from ②. The energy level diagram of the dye approaches an ideal four-level laser system in which the terminal lasing level ② corresponds to ground state dye levels with negligible thermal population, that is, states whose energy is greater than kT ($kT = 200\,\mathrm{cm}^{-1}$ at room temperature) above the ground level ①.

Besides fluorescence decay, two other modes of energy relaxation are available to a molecule in ③. These processes compete with fluorescence decay and may inhibit sustained laser oscillations. The molecule may decay directly to S_0 by nonradiative processes known as *internal conversion.* Sparse theoretical knowledge exists on this type of radiationless process in dyes. Several intramolecular quenching processes are discussed by Drexhage (1973). Representative fluorescence quantum efficiencies, that is, the ratio of fluorescence photons to absorbed photons, of laser dyes are 90%, which indicates the relative unimportance of this process for laser operation.

However, molecules in S_1 may also undergo fast nonradiative decay, of the order of 10 nsec, to the lower-lying triplet state T_1 by a process known as *intersystem crossing.* Since the triplet–singlet transition is spin forbidden, the radiative lifetime of molecules in T_1 is much longer than the fluorescence decay time from S_1 to S_0. The radiative decay $T_1 \rightarrow S_0$ is termed *phosphorescence.* Its lifetime τ_T is greater than 10^{-3} sec. Consequently, the triplet state T_1 acts as a sink for excited state molecules reducing the number of molecules that can partake in stimulated emission. In addition, absorptive transitions from T_1 to other triplet states (shown by the solid line between T_1 and T_x in Fig. 3) often occur at wavelengths that correspond to possible laser transitions. Under this condition the dye becomes more absorbing at the laser wavelengths as the population accumulates in the triplet levels and when the absorption exceeds the gain, laser action ceases.

Optical absorption between S_1 and S_2 also presents a possible source of optical loss in the dye laser. However, the internal conversion between S_2

(and higher excited states) and S_1 is usually extremely rapid, $\tau_{nr} \approx 10^{-12}$ sec, resulting in negligible optical loss.

It has been calculated that the longest laser pulse duration t_{max} consistent with $\tau_t \gg \tau_f$ is given by $t_{max} \approx 10\tau_{st}$ (Snavely, 1973), which yields $t_{max} = 0.1$ μsec. Experimental evidence has confirmed this value and demonstrates the advantage of pumping dyes with the short pulses from *Q*-switched lasers. Clearly, methods to overcome the problem created by intersystem crossing needed to be developed before CW laser operation could become a reality. After the first dye laser was constructed (Sorokin and Lankard, 1966), four years of intense effort were required before CW laser action was finally realized (Peterson *et al.*, 1970).

Excitation pulse widths less than 100 nsec were effective for pumping the first dye lasers to the limitation imposed by triplet state absorption. The initial demonstration of dye laser oscillation was accomplished with *Q*-switched ruby laser excitation. The addition of chemical additives to quench the triplet state produced longer laser pulses. Pulse durations from 100 to 500 μsec were observed with molecular oxygen (Snavely and Schafer, 1969) and cyclooctatetraene (Pappalardo *et al.*, 1970) triplet quenchers. These triplet quenchers reduced the triplet lifetimes enabling the triplet population to reach its steady state value in these time periods, which indicated that triplet absorption need not restrict CW dye laser operation.

Furumoto and Ceccon (1969) designed a small coaxial flash lamp with a fast rise time and a short pulse width that produced dye laser oscillation with a time behavior similar to that of laser-pumped dye lasers. Since then, intense dye laser pulses have been obtained with large coaxial lamps. Broad-band lasers have been constructed that deliver 32 J in 3 μsec and 110 J in 20 μsec (Snavely, 1973). Flash lamps are more efficient and less expensive than discrete-wavelength lasers for the pumping of pulsed dye lasers. However, the deposition of a large amount of energy into the dye creates serious thermal problems. The generation of a good optical-quality beam in a pulsed flash-lamp system requires careful design of the lamp, the pumping geometry, and the dye circulation and cooling system to overcome heating effects. Thermal effects have thwarted all efforts to produce CW dye laser emission in a stationary dye.

The CW dye laser avoids the triplet state bottleneck and mitigates thermal effects by the rapid flow of dye fluid through a small excitation region. Molecules reaching the triplet state are quickly removed from the optical path and replenished by a fresh supply of ground state molecules. The excitation is supplied by the TEM_{00} mode of a CW laser so that the radiation can be focused to a small spot size ~ 20 μm in the dye. The threshold power intensity for practical systems is usually about 50 kW cm^{-2} (Snavely, 1973).

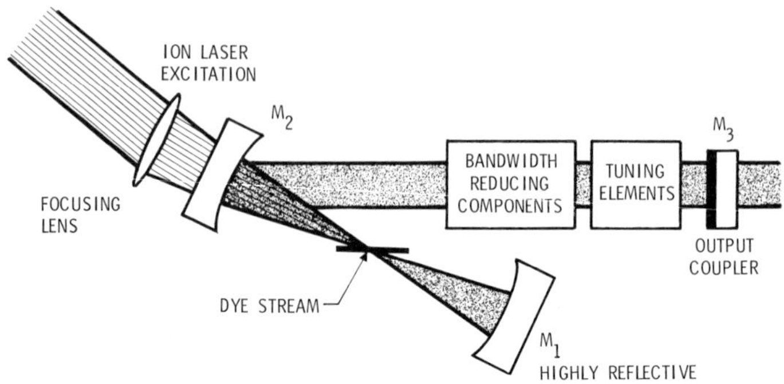

Fig. 5. Schematic representation of the resonator configuration of a CW dye laser.

A mechanical pump forces dye liquid through a nozzle and causes it to form a thin laminar sheet ($\sim$0.25 mm) that flows transversely through the excitation beam at a rate of $\sim$10 m sec^{-1}. Each dye molecule resides in the optical path for only a few microseconds.

A successful dye laser cavity design is illustrated in Fig. 5. It is formed by three mirrors and is astigmatically compensated (Kohn *et al.*, 1971). Focused laser radiation is introduced to the dye stream through a mirror M_2 that is highly transmitting at the pump wavelength and highly reflective at the dye laser wavelength. The space bounded by M_2 and M_3 provides a region where the beam is collimated so that dispersive elements can be inserted for wavelength tuning of the emission. Additional optical elements placed in the optical path can spectrally reduce the dye laser bandwidth. By proper selection of dyes, the 420–690-nm wavelength range has been spanned by a CW tunable dye laser pumped by an argon ion laser. Pumping with the strong red lines of a krypton ion laser extends the wavelength range to 900 nm. Ion laser power of up to 20 W is converted to tunable dye laser emission with conversion efficiencies of 10–50% with saturation observed above this pump level.

2. *Tuning the Dye Laser*

Tunable stimulated emission covering the 340–1200-nm wavelength region has been generated by dye lasers. The fluorescence bandwidth of individual dyes typically extends over 50–100 nm, the most efficient dyes emitting in the visible portion of the spectrum. The coumarin dyes are widely used in the blue-green region and rank among the most efficient dyes known today. Xanthene dyes span the 500–700-nm region and are generally very efficient emitters. This class of dyes boasts the most versatile of all laser dyes, rhodamine-6G. The oxayine dyes provide coherent emission across 600–800 nm while the carbocyanines have been shown to exhibit CW emission between 700 nm and 950 nm. A complete description of the molecular structure and chemical properties of laser dyes is given by Drexhage (1973). Output power versus wavelength of individual dyes pumped by flash lamps and CW lasers can be found in the literature (Marling *et al.*, 1974; Yarborough, 1974).

In the absence of wavelength-discriminating elements in the optical cavity, the dye laser will oscillate at the wavelength that maximizes the difference between the optical gain and optical losses. The laser wavelength of peak emission intensity usually occurs slightly to the long wavelength side of the fluorescence peak (see Fig. 4), and the optical bandwidth of the stimulated emission can be as large as 10 nm. Alternation of the dye concentration permits tuning the laser wavelength within a range of more than 50 nm. However, concentration tuning is rather clumsy for performing spectroscopic studies that require rapid tuning.

The use of dispersive optical components in the laser cavity provides a convenient method for wavelength tuning. In the initial demonstration of tuning, Soffer and McFarland (1967) replaced one mirror of the cavity with a diffraction grating. Then they tuned across 50 nm of the emission bandwidth of rhodamine-6G with 0.06-nm resolution. The observed power was 60% of the power recorded with the broadband mirror (60-nm resolution). Today, common tuning elements include diffraction gratings, prisms, birefringent filters, and wedge-shaped etalons. The significant optical losses of most gratings limit their use to high gain systems such as the pulsed dye laser. Metallic gratings are sometimes damaged by high-power pulses. Holographic gratings are less subject to optical damage. The small optical loss of prisms makes them appropriate dispersive components for CW systems.

Individual dyes can be tuned over 50–100 nm. The dye laser bandwidth is typically 1–3 cm^{-1} (1 $cm^{-1} \leftrightarrow$ 0.025 nm at 500 nm) when a grating or a prism tunes the laser wavelength. An intracavity etalon is required to further narrow the optical bandwidth. An etalon consists of two highly reflective surfaces separated by a fixed distance and is often composed of a slab of transparent material the faces of which are polished parallel and reflectively coated. A single etalon reduces the bandwidth to $\sim$0.1–0.3 cm^{-1}; two or more etalons can constrain the spectral width to less than 0.03 cm^{-1}. temperature stabilized etalons have restricted the laser bandwidth to a single-cavity mode. With proper control of the ambient temperature and vibrations, the emission bandwidth can be as narrow as 0.001 cm^{-1}! However, intracavity etalons introduce additional cavity losses that increase the lasing threshold and reduce the output power of the laser. Characteristics of pulsed and CW commercial dye lasers are summarized in Table II.

It is doubtful that additional dyes will be found to extend the wavelength region of dye emission much beyond the present spectral range. The short-wavelength limit is determined by photochemical decomposition that occurs in dyes at wavelengths shorter than 250 nm. The 50–100-nm wavelength

TABLE II

Characteristics of Commercial Dye Laser Systems

Excitation source	Wavelength range (nm)	Spectral width (cm^{-1})	Energy/pulse (mJ)	Pulse duration	Repetition rate (Hz)	Frequency doubled range (nm)
Flash lamp	440–700	3.0	10–100	250 nsec	10	330–440
		0.1	0.02–0.2	2 μsec	1	265–350
N_2 laser (1 MW)	360–750	0.3	0.3–1.0	8 nsec	1–50	258–360
		0.03	0.03–0.1			
Nd: YAG	530–670	1.5	10^{-3}–3×10^{-2}	40–350 nsec	3–2000	265–419
				Average power		
Ar (5 W, visible), Kr	510–900	1		0.1–1.0 W		258–300
		0.001		500 mW		
Ar (2 W, uv)	420–510	1		50–200 mW		—
		0.001		10 mW		

separation between the absorption band and the emission band of the dye (see Fig. 4) suggests that the lower-wavelength limit for dye laser oscillation will be near 300 nm. On the long-wavelength side, the loss of lasing efficiency with the decrease in energy difference between S_0 and S_1 owing to increased internal conversion is anticipated to deter efforts to extend dye laser emission further into the infrared. At the present time, it appears unlikely that dye lasers will be developed that will oscillate at wavelengths longer than 1.5 μm.

3. Mode Locking the Dye Laser

The enormous fluorescence bandwidth of dyes suggests that mode-locked dye lasers can generate high-peak power subpicosecond pulses. The bandwidth is sufficiently broad to permit tuning of the subpicosecond pulses. The subject of mode-locked dye laser has been treated by Shank and Ippen (1973). Mode locking with saturable absorbers have produced the best results. Picosecond pulses have been generated over the 430–1160-nm wavelength region and continuously from 560 to 700 nm. Pulse durations of 0.5–1.0 psec and several kilowatts of peak power have been measured for a cavity-dumped CW rhodamine-6G mode-locked laser at repetition rates up to 10^5 pps (Shank and Ippen, 1974). Peak powers of 20 MW have been recorded from pulsed mode-locked dye lasers. It will not be long before mode-locked lasers deliver 10^8–10^9 W peak power that is continuously tunable across 340–1200 nm. Although there may be little need for high peak-power picosecond pulses in the visible and near infrared regions for photoacoustic spectroscopy, these lasers are destined to play a major role in extending the wavelength coverage of tunable coherent radiation down to 100 nm by serving as the pump sources for nonlinear optical devices. These devices are the subject of the next section.

IV. Indirect Laser Sources

A. Nonlinear Optics

When radiation is incident upon a medium, a resultant polarization **P** is generated. The general form of the polarization induced by an electric field **E** is given by

$$\mathbf{P} = \chi_1 \cdot \mathbf{E} + \chi_2 : \mathbf{E} \cdot \mathbf{E} + \chi_3 \,\vdots\, \mathbf{E} \cdot \mathbf{E} \cdot \mathbf{E} + \cdots, \tag{1}$$

where χ_i is the electronic susceptibility. The first term is the usual linear susceptibility. The remaining terms constitute the nonlinear macroscopic response. The fundamental principles of nonlinear optics have been reviewed by Bloembergen (1965) and by Terhune and Maker (1968). For large field strengths in excess of 10^6 W cm^{-2}, a significant portion of the polarization may arise through the nonlinear components of **P**. The requisite radiation intensities are extremely difficult to achieve with conventional light sources, but the development of *Q*-switched and mode-locked lasers has made the production of optical intensities of 10^9 W cm^{-2} a common occurrence. Nonlinear effects can even be initiated by merely tightly focusing the radiation emitted by a 1–10-W CW laser operating in the TEM_{00} mode.

The second-order nonlinear susceptibility of materials may be utilized to produce new laser frequencies by harmonic generation, frequency mixing, or parametric amplification, thereby, extending the wavelength coverage of tunable coherent sources. Two conditions are necessary for three-wave mixing processes to proceed efficiently. Consider optical fields at three frequencies ω_1, ω_2, and ω_3 propagating in the nonlinear medium. The conditions for efficient energy transfer between the fields require

$$\omega_3 = \omega_1 \pm \omega_2, \tag{2}$$

$$\mathbf{k}_3 = \mathbf{k}_1 \pm \mathbf{k}_2, \tag{3}$$

where $\mathbf{k}_i$ is the wavevector for a wave propagating in the $\mathbf{k}_i$ direction at frequency ω_i with $|\mathbf{k}_i| = 2\pi\omega_i n_i/c$ and n_i the refractive index. These conditions may be viewed as corresponding to conservation of energy and momentum. For each photon created at ω_3, a photon at ω_1 and ω_2 must be annihilated. To understand the second condition, it is useful to imagine three waves propagating through the medium. Efficient energy exchange requires that the waves travel in phase. Equation (3) is depicted in Fig. 6. When the condition upon the phases of the waves is satisfied, the process is called "phase matched." From Fig. 6 and Eq. (2), Eq. (3) may be written as

$$\omega_3 n_3 = \omega_1 n_1 \cos\phi_1 + \omega_2 n_2 \cos\phi_2 . \tag{4}$$

It is readily seen from Eqs. (2) and (4) that, for specified orientations of the optical fields and values of ω_1, ω_2, and ω_3, constraints are imposed upon the refractive indices n_i. The refractive index of a medium varies with fre-

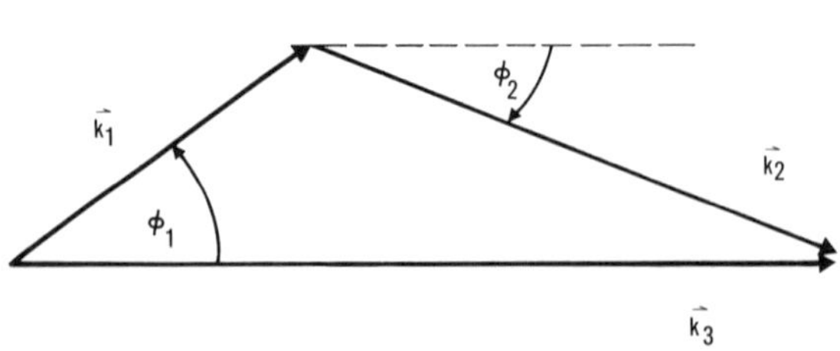

Fig. 6. Conservation of momentum for a three-wave mixing process. Collinear phase-matching is the case for which $\phi_1 = \phi_2 = 0$.

quency in a prescribed manner; therefore, the phase-matching requirement is not generally satisfied.

Phase matching is readily accommodated by the natural birefringence of certain anisotropic materials. Birefringence refers to the condition in which the refractive index of an optical beam passing though a medium depends upon the direction of propagation and polarization. The phenomenon is well understood, and its theory is rigorously treated by many authors (Nye, 1957; Born and Wolf, 1965). For our purposes, it suffices to note that numerous materials, in particular, noncentrosymmetric crystals that exhibit negative birefringence, have been fabricated for the nonlinear generation of light.

B. *Second Harmonic Generation*

Second harmonic generation refers to the three-wave mixing process in which the frequency of two of the three fields is the same, that is, $\omega_1 = \omega_2 = \omega$. Maximum second harmonic generation occurs when

$$\omega_3 = \omega_1 + \omega_2 = 2\omega, \tag{5}$$

$$\mathbf{k}_{2\omega} = 2\mathbf{k}_{\omega}, \tag{6}$$

where $\mathbf{k}_{2\omega}$ and $\mathbf{k}_{\omega}$ are the wavevectors at the second harmonic and fundamental frequency, respectively. For Gaussian beams, the conversion efficiency η of power at the fundamental frequency P_{ω} to the power at the second harmonic $P_{2\omega}$ is given by

$$\eta = \frac{P_{2\omega}}{P_{\omega}} = 2\left(\frac{\mu_0}{\varepsilon_0}\right)^{3/2} \omega^2 \frac{d^2}{n^3} L^2 \frac{P_{\omega}}{A} \sin^2\theta \frac{\sin^2(\Delta k L/2)}{(\Delta k L/2)^2}, \tag{7}$$

where μ_0 is the free space permeability, ε_0 the free space permittivity, d the second-order nonlinear optical coefficient, L the crystal length, n the refractive index, A the cross-sectional beam area, Δk the phase mismatch $= |\mathbf{k}_{2\omega} - \mathbf{k}_{\omega}|$, and θ the angle between the direction of propagation of fundamental beam and the crystal axis. It can be seen from Eq. (7) that maximum efficiency requires: large optical intensities, long crystals, strong nonlinear interaction as measured by d, perfect phase matching, and 90° phase matching, that is, $\theta = 90°$.

Second harmonic generation was first observed by Franken *et al.* (1961). The radiation from a pulsed ruby laser oscillating at 694 nm was focused into a quartz crystal where it was converted to the second harmonic wavelength at 347 nm. It is common practice to achieve 30% conversion efficiency for the production of the second harmonic of ruby (Kato *et al.*, 1974).

Associated with the exploitation of the natural birefringence of certain uniaxial crystals for harmonic generation is the problem of beam "walk-off"

(Born and Wolf, 1965). In a uniaxial crystal, the direction of power flow, as given by the Poynting vector, does not generally coincide with the beam propagation direction. After traveling a certain distance inside the crystal, the fundamental beam and the second harmonic beam will physically separate, thus terminating the energy exchange between the optical fields and limiting the generation of second harmonic power. A locus of orientations exists for which the Poynting vector and the direction of propagation are collinear. It is the plane perpendicular to the optic axis. When the phase-matching requirement for second harmonic generation is satisfied in this plane, the process is referred to as "90° phase matching" and results in the elimination of the walk-off problem. The desirable material properties for any three-wave mixing process that arises from the second-order nonlinear susceptibility χ_2 are

(1) transparency at ω_1, ω_2, and ω_3;
(2) large d;
(3) large birefringence;
(4) long crystal lengths;
(5) good optical homogeneity;
(6) high threshold for damage by intense optical fields;
(7) good chemical stability.

The more common materials for second-order nonlinear generation of tunable radiation appear in Table III (Adhav and Orszag, 1974). The conversion efficiencies shown in Table III have been recorded for a 15 mm long crystal and a 100-MW cm^{-2} pulse from a Q-switched laser. Higher conversion efficiencies are possible with longer crystals and more intense beams.

At the present time, CDA is the most efficient material for doubling the frequency of the Nd:YAG laser. The fundamental of the ruby laser is optimally converted to second harmonic in RDA. The emission of dye lasers between 570 and 610 nm can be 90° phase matched to generate second harmonic in ADA. The ADP and KDP crystals efficiently convert radiation between 500 and 550 nm to the shortest wavelengths achievable today by second-harmonic generations, 250–275 nm. As such, they are useful for doubling several wavelengths of the argon ion laser and the second harmonic of Nd:YAG at 532 nm to radiation at 266 nm. Generation of wavelengths shorter than 250 nm by second harmonic generation awaits new developments in the field of crystal technology.

Hygroscopic crystals are easy to grow but exhibit poor chemical stability. The polished faces may be fogged if hand held or breathed upon. The nonhygroscopic crystals are easy to handle but difficult to grow and thus more

TABLE III

Nonlinear Crystal Parameters

Crystal					
Common name	Chemical formula	Hygroscopic	Transparency region (nm)	Typical second harmonic conversion efficiency (%) (1.06 → 0.53 μm)	Damage threshold (MW cm^{-2})
ADP	$NH_4H_2PO_4$	Yes	200–1600	5	100
KDP	KH_2PO_4	Yes	200–1600	8	100
ADA	$NH_4H_2AsO_4$	Yes	260–1600	—	300
RDA	RbH_2AsO_4	Yes	260–1600	25	300
CDA	CsH_2AsO_4	Yes	260–1600	30	300
LFM	$LiCOOH \cdot H_2O$	Yes	250–1200	14	200
Lithium iodate	$LiIO_3$	Yes	320–4000	8	100
Lithium niobate	$LiNbO_3$	No	420–4200	6	50
Bananas	$BaNaNb_5O_{15}$	No	460–1060	12	5

expensive. The above crystals work most efficiently under restricted temperature and orientation over a relatively narrow wavelength range. The nonlinear coefficients of these and other crystals have been tabulated by Singh (1971). Nonlinear crystals are commercially available and routinely sold as accessories that fit directly into laser systems.

Pulsed tunable lasers deliver sufficient peak power to allow second harmonic generation of ultraviolet radiation with ~20% conversion efficiency. However, CW tunable dye lasers generate approximately 1 W of power; therefore, less than 1% of the fundamental power is converted to second harmonic power on a CW basis. The conversion efficiency for low power lasers can be remarkably improved by the placement of the nonlinear crystal inside the cavity, thereby taking advantage of the intense optical fields that exist inside the laser resonator. It has been theoretically shown that 100% of the power at the output of an optimally coupled laser can be converted to second harmonic power by internal second harmonic generation (Yariv, 1971). Geusic *et al.* (1968) demonstrated 100% conversion efficiency of Nd:YAG emission to power at 0.53 μm by internal doubling in $BaNaNb_3O_{15}$. Intracavity second harmonic generation provides the feature of internal self-adjustment with respect to the level at which the laser is operated. Once the system is adjusted to the optimum coupling value for second harmonic generation, it retains this value for all levels of excitation or laser gain.

Intracavity second harmonic generation has produced tunable ultraviolet radiation. Wallace (1971) demonstrated tuning between 261 and 315 nm in a commercial system. An ADP crystal was placed inside the cavity of a tunable dye laser pumped by a repetitively pulsed, frequency-doubled Nd:YAG laser. Gabel and Hercher (1972) placed a nonlinear crystal inside the cavity of a rhodamine-6G laser pumped by a 1-W CW argon ion laser and produced tunable second harmonic radiation across the 290–315-nm spectral range.

Commercial units routinely frequency double the output of tunable dye lasers and, thereby, generate tunable coherent radiation to wavelengths as short as 250 nm.

C. *Sum-Frequency Mixers*

Even though ADP and KDP crystals remain transparent to 200 nm, direct frequency doubling to the short-wavelength edge of their transparency region is not possible because of insufficient crystal birefringence to allow proper phasematching, that is, the refractive indices of ADP and KDP do not permit the simultaneous solution of Eqs. (5) and (6) for second harmonic wavelengths less than 250 nm. *Frequency mixing* is not subject to this constraint and provides a method for the generation of tunable coherent emission between 200 and 250 nm.

Frequency mixing refers to the second-order nonlinear process in which two optical fields at different frequencies are incident upon the nonlinear crystal. Inside the crystal, the fields combine to create radiation at the sum or difference frequency. The energy and momentum conditions that rule the mixing process are given by Eqs. (2) and (3). When $\omega_3 = \omega_1 - \omega_2$, the process is termed *difference-frequency mixing* and is used in the production of tunable infrared radiation. As such, no further reference to this process will be made here. The complementary process, that is, $\omega_3 = \omega_1 + \omega_2$, is named *sum-frequency mixing* and provides a technique for the generation of tunable coherent ultraviolet energy. In general, the process requires input radiation from two lasers.

Massey (1974) has shown that when one of the input frequencies is close to ω_3, the energy and momentum conditions can be satisfied in ADP for the three-wave sum-mixing process that produces radiation over the wavelength range 200–250 nm. The fourth harmonic of a Q-switched, repetitively pulsed Nd:YAG laser was mixed with the fundamental at 1.06 μm in ADP, and 60% of the power at 266 nm was converted to coherent ultraviolet light at 212.8 nm. Sum-frequency mixing of intense Nd:YAG radiation with the second-harmonic output of tunable dye lasers in the 250–300-nm range will provide tunable radiation in the 200–250 nm spectral region.

D. *Optical Parametric Oscillators*

Such reactive media as nonlinear crystals provide conditions whereby power may be exchanged between optical fields at different frequencies. The optical parametric oscillator uses nonlinear crystals to generate tunable coherent radiation. It is of considerable practical importance for the production of tunable coherent light in the visible and infrared spectral regions. Commercial systems can be purchased. The tuning range of a single parametric oscillator (~ 1 μm) may be more extensive than that of individual laser dyes (~ 0.1 μm). However, its bandwidth is typically greater than 1 cm^{-1}, whereas, the spectral width of a dye laser can be less than 0.01 cm^{-1}. The parametric oscillator has been reviewed by Harris (1969).

The optical parametric oscillator consists of a nonlinear crystal inside an optical cavity. A pumping field irradiates the crystal. Inside the crystal, photons at the pump frequency ω_p decay into photons at the "signal" frequency ω_s and "idler" frequency ω_i. The terminology is carried over from microwave parametric oscillators, and ω_i, ω_s, and ω_p correspond to ω_1, ω_2, and ω_3, respectively, of Section IV.A. The optical cavity is bounded by mirrors that are resonant at either ω_i or ω_s or both. The conservation conditions for optimum generation of parametric power are given by $\omega_p = \omega_i + \omega_s$, and $\mathbf{k}_p = \mathbf{k}_i + \mathbf{k}_s$. The wavevectors $\mathbf{k}_p$, $\mathbf{k}_i$, and $\mathbf{k}_s$ of the pump, idler, and signal wave, respectively, are proportional to the refractive index of the crystal at their respective frequencies. Once the pump frequency is selected, the frequency of the signal and idler waves will tune to satisfy the conservation requirements upon variation of the refractive index. The refractive index of the nonlinear crystal is dependent upon temperature, the angle between the pump field and cavity axis, application of electrical field, and pressure. Common parametric oscillators are temperature and angle tuned.

The amount of wavevector mismatch that can be tolerated before the phases of the three waves are out of synchronism determines the bandwidth of the parametric oscillator. The bandwidth $\Delta\nu$ (cm^{-1}) is given by

$$|\Delta\nu| = 1/cbL, \tag{8}$$

where b is a dispersive constant given by

$$b = (\partial k_i/\partial\omega_i) - (\partial k_s/\partial\omega_s). \tag{9}$$

The linewidth of a parametric oscillate that uses a 1-cm KDP crystal is $\sim 40\,cm^{-1}$; for a 1-cm crystal of $LiNbO_3$, $\Delta\nu \sim 5\,cm^{-1}$. The linewidth of the oscillator is the largest near degeneracy, that is, for $\omega_i \approx \omega_s$.

Falk and Murray (1969) constructed an angle-tuned parametric oscillator that was pumped by a pulsed ruby laser and was tuned from 1.05 to 1.20 μm and 1.64–2.05 μm. Energy conversion of 45% was achieved. Wallace (1972) produced a temperature-tuned $LiNbO_3$ parametric oscillator that was continuously tunable over the wavelength range from 0.76 to 2.6 μm. It was pumped by a flash-lamp-pumped tunable rhodamine-6G dye laser. The commercial version of this oscillator generates up to 0.5 mJ per pulse and its bandwidth is normally 2–2.5 cm^{-1} (Chromatix, 1975).

E. Third-Order Nonlinear Generation of Tunable Vacuum Ultraviolet Radiation

Because of crystal absorption, the shortest wavelength that can be derived from nonlinear birefringent crystals is limited to 200 nm. Third-order nonlinear optical processes in metal vapors and inert gases have produced tunable coherent radiation below 200 nm in a spectral region known as the vacuum ultraviolet (vuv). The inversion symmetry of gases causes χ_2 to vanish; thus, the production of radiation by second-order nonlinear optical processes is forbidden in the gaseous phase. The third-order process takes place through the third term in the series expansion for the electronic polarization as a function of electric field (see Eq. 1); namely, $\mathbf{P}_3 = \chi_3\mathbf{E}^3$ where $\mathbf{P}_3$ is the third-order polarization and χ_3 is the third-order nonlinear susceptibility. Under phase-matched conditions, optical fields at three frequencies ω_1, ω_2, and ω_3 combine to produce radiation at a fourth frequency ω_4 given by $\omega_1 \pm \omega_2 \pm \omega_3 = \omega_4$. Two or more frequencies may be identical. Third-harmonic generation corresponds to the case in which $\omega_1 = \omega_2 = \omega_3$. The four-wave interaction most useful for the generation of vacuum ultraviolet radiation occurs for $\omega_1 + \omega_2 + \omega_3 = \omega_4$.

The media requirements for efficient production of vuv power are (1) vuv transparency, (2) phase matching, and (3) strong third-order nonlinear susceptibility. Ward and New (1969) recognized that alkali metal vapors and inert gases fulfill the requirements for efficient four-wave mixing. Calculations by Miles and Harris (1973) indicate that large conversion efficiencies are feasible for vapor mixtures pumped by picosecond laser pulses with peak power of 10^8–10^9 W. By third-harmonic generation in sodium and rubidium, 30 psec pulses of radiation at 1064 nm have been converted to 355 nm with an energy conversion efficiency of 10% (Bloom *et al.*, 1975a). Xenon was used as the buffer gas to compensate the anomalous dispersion of the metal vapor, thus the refractive index of the mixture was the same at the fundamental frequency and the third harmony frequency.

Recent experiments with mixed metal vapors have overcome the limitations on efficient third-harmonic generation owing to the high buffer gas pressure required for phase-matching (Bloom *et al.*, 1975b). Phase-matched frequency tripling was observed in a homogenous mixture of magnesium and sodium vapor at 11 and 5 torr, respectively. Lower phase-matching pressures may ultimately lead to higher conversion efficiencies.

The large resonant enhancement in χ_3 when one of the input frequencies is tuned to a double-quantum transition of the metal vapor has been exploited by Hodgson *et al.* (1974) to produce tunable vuv radiation in strontium vapor. The coherent, narrow bandwidth (0.1-cm^{-1}) output was continuously tunable from 157 to 196 nm. With other vapors, such as mercury, magnesium, or sodium, this method should generate tunable radiation from 80 to 230 nm.

Phase matching requires the temperature of the gas to be uniform over the interaction length to 0.5°C and is more difficult to achieve with metal vapors than inert gases. Kung *et al.* (1973) reported 3% conversion efficiency for the generation of 118-nm radiation in a phase-matched mixture of xenon and argon gases. Later work generated tunable vuv radiation over the 118–147-nm and 163–195-nm spectral region by the third-order nonlinear mixing of a broadly tunable visible parametric oscillator and the fourth harmonic of a mode-locked Nd:YAG laser in xenon gas (Kung, 1974). The mixing process was not properly phase matched, and the measured output power was typically 1 W in a 20-psec pulse. With pump radiation produced by harmonic generation of tunable dye laser emission and fixed wavelength lasers, the general technique of phase-matched harmonic generation in mixtures of inert gases will provide tunable coherent radiation to wavelengths as short as 50 nm.

V. Summary

Continuous sources of radiation throughout the 0.1–1.0-μm spectral region have been reviewed. The tungsten lamp and the xenon arc generate intense incoherent emission across the visible and near infrared portion of the spectrum. High-pressure mercury arcs emit strongly at numerous discrete wavelengths between 230 and 650 nm. Several sources radiate moderate amounts of incoherent energy in various bands in the 110–230 nm region.

Laser emission is characterized by extreme spectral purity and excellent collimation. These properties combine to make laser sources ideal for high-resolution spectroscopic work. A summary of the wavelength coverage of important broadly tunable laser sources appears in Fig. 7. With the exception of the last method, all of the components of the laser sources that appear in Fig. 7 are commercial products. Dye laser emission has been continuously tuned across the 340–1200-nm spectral region. The optical parametric oscillator delivers broadly tunable visible and infrared radiation. Nonlinear

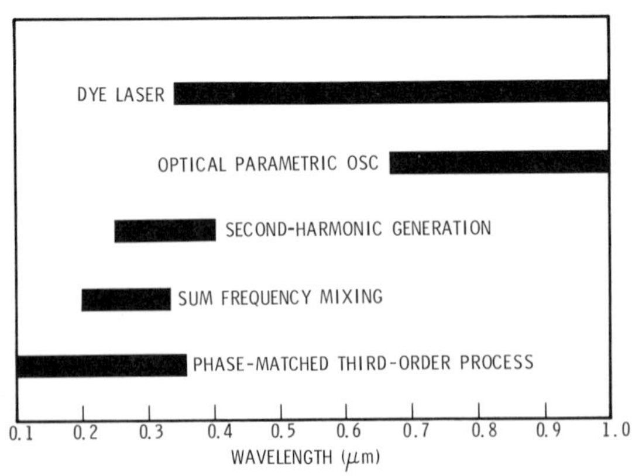

Fig. 7. Methods for the generation of tunable coherent emission between 0.1 and 1.0 μm, and the wavelength region where each is important.

optical processes provide tunable coherent emission at shorter than visible wavelengths. Second-harmonic generation of dye laser emission has produced coherent radiation down to 250 nm. Sum-frequency mixing of two input lasers in nonlinear media extends the wavelength coverage to 200 nm. The 100–200-nm spectral region has been spanned in the laboratory by phase-matched third-order nonlinear generation in metal vapors and inert gases.

Acknowledgments

The author wishes to express his gratitude to Drs. T. S. Hartwick, J. P. Hurrell, P. F. Jones, R. L. Kohn, and J. E. Wessel for their careful reading of this manuscript and their helpful comments.

References

Adhav, R. S., and Orszag, M. (1974). *Electro–Opt. Syst. Des.* **6**, 20–24.

Billings, B. H. (1972). *In* "American Institute of Physics Handbook," 3rd ed., pp. 6-216–6-127. McGraw-Hill, New York.

Bloembergen, N. (1965). "Nonlinear Optics." Benjamin, New York.

Bloom, D. M., Bekkers, G. W., Young, J. F., and Harris, S. E. (1975a). *Appl. Phys. Lett.* **26**, 687–689.

Bloom, D. M., Young, J. F., and Harris, S. E. (1975b). *Appl. Phys. Lett.* **27**, 390–392.

Born, M., and Wolf, E. (1965). "Principles of Optics." Pergamon, Oxford.

Bridges, W. B., and Chester, A. N. (1971). *In* "Handbook of Lasers" (R. J. Pressley, ed.), pp. 250–251. Chemical Rubber Co., Cleveland, Ohio.

Boyd, G. D., and Gordon, J. P. (1961). *Bell Syst. Tech. J.* **40**, 489–508.

Boyd, G. D., and Kogelnik, H. (1962). *Bell Syst. Tech. J.* **41**, 1347–1369.

Calvert, J. G., and Pitts, J. N., Jr. (1966). "Photochemistry." Wiley, New York.

Cann, M. W. P. (1969). *Appl. Opt.* **8**, 1645–1661.

Chang, W. S. (1969). "Principles of Quantum Electronics." Addison-Wesley, Reading, Massachusetts.

Chromatix (1975). CMX-4/IR Preliminary Data Sheet.

DeMaria, A. J., Glenn, W. H., Jr., Brienza, M. J., and Mack, M. E. (1969). *Proc. IEEE* **57**, 2–25.

Deutsch, T. F. (1971). *In* "Handbook of Lasers" (R. J. Pressley, ed.), pp. 350–354. Chemical Rubber Co., Cleveland, Ohio.

Drexhage, K. H. (1973). *In* "Dye Lasers" (F. P. Schafer, ed.), pp. 144–193. Springer-Verlag, Berlin and New York

Falk, J., and Murray, J. E. (1969). *Appl. Phys. Lett.* **14**, 254–247.

Fox, A. G., and Li, T. (1960). *Proc. IRE* **48**, 1904–1905.

Franken, P. A., Hill, A. E., Peters, C. W., and Weinreich, G. (1961). *Phys. Rev. Lett.* **7**, 118–119.

Furumoto, H., and Ceccon, H. (1969). *J Appl. Phys.* **40**, 4204–4205.

Gabel, C., and Hercher, M. (1972). *IEEE J. Quantum Electron.* **8**, 850–851.

Gerry, E. T. (1965). *Appl. Phys. Lett.* **7**, 6–8.

Geusic, J. E., Levinstein, H. J., Singh, S., Smith, R. G., and VanUitert, L. G. (1968). *Appl. Phys. Lett.* **12**, 306–308.

Godard, B. (1974). *IEEE J. Quantum.* **10**, 147–153.

Harris, S. E. (1969). *Proc. IEEE* **57**, 2096–2113.
Heard, H. G. (1963). *Nature* (*London*) **200**, 667.
Hellwarth R. W. (1966). *In* "Lasers" (A. K. Levine, ed.), Vol. I, pp. 253–294. Dekker, New York.
Hodgson, R. T., Sorokin, P. P., and Wynne, J. J. (1974). *Phys. Rev. Lett.* **32**, 343–346.
Kasper, J. (1972). *In* "American Institute of Physics Handbook" (D. E. Gray, ed.), 3rd ed., pp. 6–198. McGraw Hill, New York.
Kato, K., Alcock, A. J., and Richardson, M. C. (1974). *Opt. Commun.* **11**, 5–7.
Kogelnik, H. and Li, T. (1966). *Proc. IEEE* **54**, 1312–1329.
Kohn, R. L., Shank, C. V., Ippen, E. P., and Dienes, A. (1971). *Opt. Commun.* **3**, 177–178.
Koller, L. R. (1965). "Ultraviolet Radiation," 2nd ed. Wiley. New York.
Kung, A. H. (1974). *Appl. Phys. Lett.* **25**, 653–654.
Kung, A. H., Young, J. F., and Harris, S. E. (1973). *Appl. Phys. Lett.* **22**, 301–302.
Lengyel, B. A. (1966). "Introduction to Laser Physics." Wiley, New York.
Leonard, D. A. (1965). *Appl. Phys. Lett.* **7**, 4–6.
Maiman, T. H. (1960). *Nature* (*London*) **187**, 493–494.
Marling, J. B., Hawley, J. G., Liston, E. M., and Grant, W. B. (1974). *Appl. Opt.* **13**, 2317–2320.
Massey, G. A. (1974). *Appl. Phys. Lett.* **24**, 371–373.
Maydan, D., and Chesler, R. B. (1971). *J. Appl. Phys.* **42**, 1031–1034.
McNesby, J. R., Braun, W., and Ball, J. (1971). *In* "Creation and Detection of the Excited State" (A. A. Lamola, ed.), Vol. I, Part B, pp. 503–582. Dekker, New York.
Miles, R. B., and Harris, S. E. (1973). *IEEE J. Quantum Electron.* **9**, 470–484.
Nye, J. F. (1957). "Physical Properties of Crystals." Oxford Univ. Press (Clarendon), London and New York.
Pantell, R. H., and Puthoff, H. E. (1969). "Fundamentals of Quantum Electronics." Wiley, New York.
Pappalardo, R., Samelson, H., and Lempiciki, A. (1970). *IEEE J. Quantum Electron.* **6**, 716–725.
Peterson, O. G., Tuccio, S. A., and Snavely, B. B. (1970). *Appl. Phys. Lett.* **17**, 245–247.
Pressley, R. J. (ed.) (1971). "Handbook of Lasers." Chemical Rubber Co., Cleveland, Ohio.
Samson, J. A. R. (1967). "Techniques for Vacuum Ultraviolet Spectroscopy." Wiley, New York.
Shank, C. V., and Ippen, E. P. (1973). *In* "Dye Lasers" (F. P. Schafer, ed.), pp. 121–143. Springer-Verlag, Berlin and New York.
Shank, C. V., and Ippen, E. P. (1974). *Appl. Phys. Lett.* **24**, 373–375.
Shipman, J. D., Jr., and Kolb, A. C. (1966) *IEEE J. Quantum Electron.* **2**, 298.
Siegman, A. E. (1971). "An Introduction to Lasers and Masers." McGraw-Hill, New York.
Singh, S. (1971). *In* "Handbook of Lasers" (R. J. Pressely, ed), pp. 489–525. Chemical Rubber Co., Cleveland, Ohio.
Smith, P. W. (1970). *Proc IEEE* **58**, 1342–1357.
Smith, W. V., and Sorokin, P. P. (1966). "The Laser." McGraw-Hill, New York.
Snavely, B. B. (1973). *In* "Dye Lasers" (F. P. Schafer, ed.), pp. 86–120. Springer-Verlag, Berlin and New York.
Snavely, B. B., and Schafer F. P. (1969). *Phys. Lett.* **28**, 728–729.
Soffer, B. H., and McFarland, B. B. (1967). *Appl. Phys. Lett.* **10**, 266–267.
Sorokin, P. P., and Lankard, J. R. (1966). *IBM J. Res. Dev.* **10**, 162–163.
Suchard, S. N., Galvan, L., and Sutton, D. G. (1975). *Appl. Phys. Lett.* **26**, 521–523.
Tanaka, Y. (1955). *J. Opt. Soc. Am.* **45**, 710–713.
Terhune, R. W., and Maker, P. D. (1968). *In* "Laser" (A. K. Levine, ed.), Vol. II, pp. 295–372. Dekker, New York.
Wallace, R. W. (1971). *Opt. Commun.* **4**, 316–318.
Wallace, R. W. (1972). *IEEE J. Quantum Electron.* **8**, 819–820.
Ward, J. F., and New, G. H. C. (1969). *Phys. Rev.* **185**, 57–72.

Wilkinson, P. G. (1955). *J. Opt. Soc. Am.* **45**, 1044–1046.
Wilkinson, P. G., and Tanaka, Y. (1955). *J. Opt. Soc. Am.* **45**, 344–349.
Wu, F. Y., Grove, R. E., and Ezekiel, S. (1974). *Appl. Phys. Lett.* **25**, 73–75.
Yarborough, J. M. (1974). *Appl. Phys. Lett.* **24**, 629–630.
Yariv, A. (1967). "Quantum Electronics." Wiley, New York.
Yariv, A. (1971). "Introduction to Optical Electronics." Holt, New York.

5

Tunable Infrared Laser Sources for Optoacoustic Spectroscopy

P. L. Kelley

Lincoln Laboratory
Massachusetts Institute of Technology
Lexington, Massachusetts

I. Introduction

This chapter concerns tunable infrared lasers that can be used in optoacoustic spectroscopy. We confine our attention to the 1–30-μm region, a wavelength interval containing many important molecular absorption bands. Figure 1 shows an example of a vibration–rotation absorption spectrum obtained using a tunable laser source together with an optoacoustic detector (Pike *et al.*, 1972).

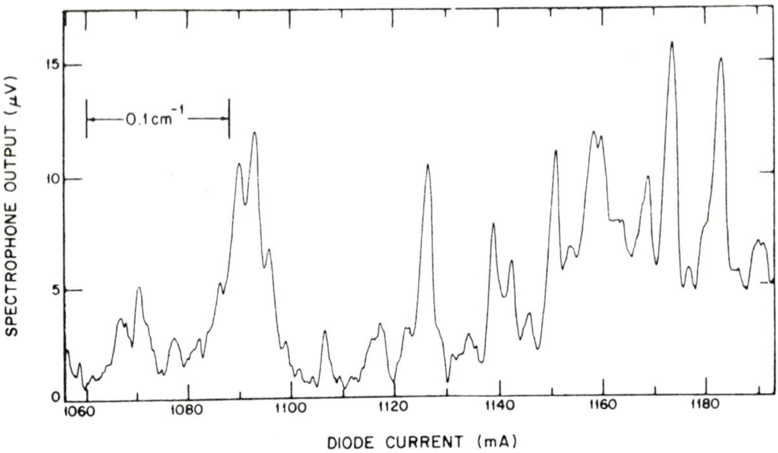

Fig. 1. Optoacoustic spectroscopy of ethylene (C_2H_4) near 10.6 μm. The optoacoustic cell (spectrophone) output signal is shown as a function of current through a $Pb_{0.88}Sn_{0.12}Te$ diode laser. As current is changed the output wavelength of the laser changes and a spectrum is obtained of a portion of the Q branch of the ν_7 band of C_2H_4. (From Pike *et al.*, 1972.)

The distinction between tunable and fixed-frequency lasers is not as clear as one might first believe: A reasonable definition of a broadly tunable laser is one that is continuously tunable over 1 cm^{-1} or more and has a total tuning range of hundreds of reciprocal centimeters; a fixed-frequency laser is one that cannot be tuned over more than a small fraction of a reciprocal centimeter (e.g., $\sim$0.002 cm^{-1} for a low-pressure CO_2 laser transition). A fixed-frequency laser may often be step tuned from one transition to another, but the resulting total continuous coverage is typically only 0.1 cm^{-1} over a 100-cm^{-1} interval, and for this reason these lasers seldom match perfectly with the absorption lines of molecules, and sensitivity is therefore reduced. Lasers that fall into the intermediate class between broadly tunable and fixed-frequency lasers include rare earth lasers and the magnetically tuned He–Ne laser.

Broadly tunable lasers generally have some method of controlling the gross-tuning as well as the fine-tuning characteristics of the laser either by prechoice of the wavelength range (changing composition in the case of semiconductor lasers, changing dyes in dye lasers) or by changing the operating conditions (tuning to the center of the transition by temperature or magnetic or electric field, or controlling the wavelength dependence of the cavity Q with prisms or gratings). Fine control of the laser wavelength generally involves tuning a single-cavity mode with an etalon or by small changes in temperature, magnetic field, or electric field. As individual tunable lasers are discussed, the tuning techniques are described where appropriate.

Lasers can show a considerable variation in linewidth, depending on the geometry and number of elements in the cavity as well as the breadth of the gain peak. Since narrowing the linewidth often involves a considerable complication to the system, it is not advisable to narrow the output beyond some reasonable fraction of the width of a spectral line (e.g., 1/10 for high specificity and 1/2 for high sensitivity). At atmospheric pressure spectral lines are typically 0.1 cm^{-1} wide (full width half maximum), while in the low-pressure Doppler limited regime they are typically of the order of 0.002 cm^{-1} at wavelengths near 10 μm. The effective laser linewidths (including frequency jitter) should, therefore, be narrower than 3 GHz and 60 MHz, respectively.

Power levels are also important (both peak and average) as well as pulse duration. Fixed-frequency lasers are generally better than tunable lasers in these characteristics, although there are exceptions. For detecting small concentrations of gas substantial cw powers may be required. At atmospheric pressure and a 1-sec integration time a well designed spectrophone can detect in the parts per million (ppm) range with a milliwatt of average power and in the parts per billion range (ppb) with a watt of average power assuming a moderate infrared absorption cross section of 10^{-18} cm^2. Where sensitivity is not a serious problem, considerably smaller average powers may be used.

Table I gives some of the parameters for several typical laser systems that are discussed in this chapter.

TABLE I

Some Tunable Infrared Lasers

		Typical power (W)	
Tunable source	Wavelength region (μm)	CW	Pulsed
Difference frequency generators	1–30	10^{-6} (2–4 μm)	10^5
Optical parametric oscillators	1–3.5 ($LiNbO_3$) 1.2–8.5 (Ag_3AsS_3) 8–12 (CdSe)	—	10^5
Semiconductor diode lasers	1–30	10^{-3}	10
Spin-flip Raman lasers	3 (HF pumped) 5–6 (CO pumped) 9–14 (CO_2 pumped)	1 (5 μm)	10^3
High-pressure gas lasers	9–11 (CO_2)	—	10^5

II. Optical Mixers and Parametric Oscillators

Optical mixers and optical parametric oscillators most often involve the use of a nonlinear crystal. In many noncentrosymmetric crystals a significantly large dielectric polarization, quadratic in the strength of an applied laser field, can be induced. The nonlinear polarization can radiate, permitting harmonic generation or, when fields at two frequencies are applied, sum- and difference-frequency generation. For a substantial buildup of radiation at these new frequencies one must satisfy the "phase matching" condition: the wave vectors (**k**) of the applied fields and the generated field have the same relationship as their frequencies (ω), so that, for difference-frequency generation where $\omega_1 - \omega_2 = \omega_3$, then $\mathbf{k}_1 - \mathbf{k}_2 = \mathbf{k}_3$. The usual way to satisfy this condition for collinear propagation is to select the propagation direction and/or crystal temperature so that crystal birefringence offsets the effects of dispersion. The difference-frequency output power is given by

$$P_3 = \frac{(8\pi\omega_3 \, dl)^2}{n_1 n_2 n_3 A c^3} P_1 P_2,$$

where d is the nonlinear coefficient, l the crystal length, n the index of refraction, A the focal area, and c the velocity of light. For the case of $LiNbO_3$ and typical spot sizes and crystal lengths the above expression reduces to

$$P_3 \approx 2 \times 10^{-4} P_1 P_2,$$

where the powers are in watts.

If one has a tunable laser in one region of the spectrum, one can get tunable outputs at longer wavelengths with difference-frequency generation. Dye lasers are often the original tunable source. In an early experiment pulsed difference-frequency generation from the output of a ruby laser plus a dye laser mixed in $LiNbO_3$ achieved 6 kW of infrared power tunable between 3.1 and 4.5 μm, with a bandwidth of about 10 wavenumbers (Dewey and Hocker, 1971). In this wavelength range further development of pulsed-laser mixing techniques has been carried out including increasing average power and obtaining narrow linewidths (Meltzer and Goldberg, 1972; Bethea, 1973; Goldberg, 1975). Doubled Nd:YAG radiation has been used as the dye laser pump and $LiIO_3$ as an alternative nonlinear material. Mixing in $LiNbO_3$ the output of a stable dye laser with that of a stable argon laser has yielded in the 2–4-μm region CW output powers of about 1 μW covering a tuning range of over one hundred wavenumbers with a spectral precision and reproducibility of 15 MHz (Pine, 1974, 1976). Figure 2 shows a schematic of this system.

Longer wavelengths can be generated by mixing in other crystals: proustite (Ag_3AsS_3) (Hanna *et al.*, 1971; Decker and Tiltel, 1973) is usable to about 12 μm and CdSe (Bhar *et al.*, 1972; Hanna *et al.*, 1974) and Te (Bridges *et al.*,

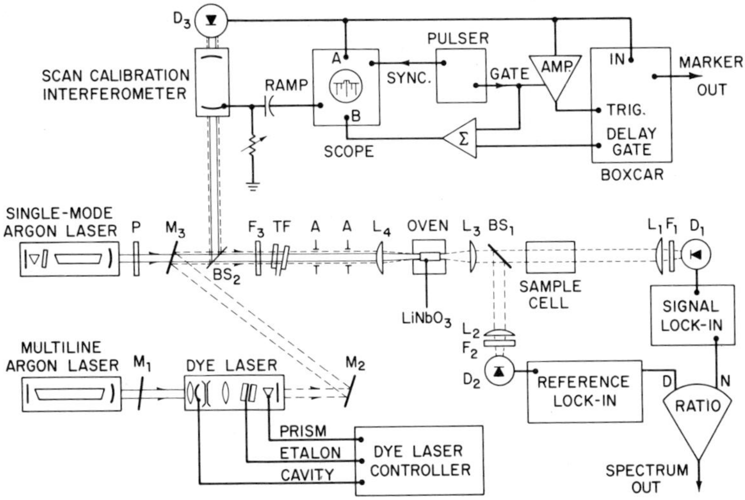

Fig. 2. Schematic of visible-to-infrared difference-frequency spectrometer system. (From Pine, 1976.)

1975) to about 25 μm. Also under development (Chemla *et al.*, 1971; Byer *et al.*, 1971; Boyd *et al.*, 1972a, b; Kildal and Mikkelsen, 1973, 1974; Kildal *et al.*, 1976) as nonlinear materials in the infrared are I–III–VI_2 and II–IV–V_2 compounds with the chalcopyrite structure (e.g., $AgGaS_2$, $AgGaSe_2$, $CdGeAs_2$, and $ZnGeP_2$). These phase-matchable materials can be used for sum- and difference-frequency generation as well as in parametric oscillators. Mixing in $AgGaS_2$ of radiation from dye lasers have given tunable output in the 4–18-μm range. Table II summarizes some of the properties of these materials (Kildal *et al.*, 1976).

TABLE II

Properties of Middle Infrared Nonlinear Materials

Material	Index of refraction n	Nonlinear coefficient $d \times 10^{22}$ (As V^{-2})	Nonlinear figure of merit d^2/n^3 (relative units)	Laser-induced surface damage threshold at 10 μm (J cm^{-2})	Transmission range (μm)
Te	4.9	57	405	—	4–25.0
$CdGeAs_2$	3.6	26	215	5–6	2.5–17.5
$ZnGeP_2$	3.1	6.6	22	—	0.7–12.0
GaSe	2.6	5.0	21	—	0.65–18.0
$AgGaSe_2$	2.6	2.9	7	1.5–3.0	0.7–19.0
$AgGaS_2$	2.4	1.1	1.2	3–3.5	0.5–12.5
Ag_3AsS_3	2.6	1.1	1	1.9	0.64–13.0

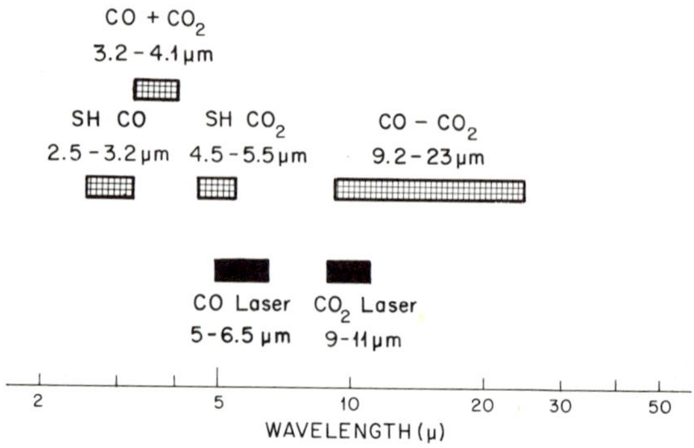

Fig. 3. Stepwise tunable coverage obtainable in the middle infrared by sum- and difference-frequency generation using CO and CO_2 lasers. (From Mooradian, 1974.)

Step tuning with nearly continuous coverage of many regions of the middle infrared can be obtained by sum- and difference-frequency generation using the outputs of CO_2, N_2O, CO, and HF/DF lasers. Figure 3 shows the coverage that can be obtained in this fashion using CO and CO_2 lasers (Mooradian, 1974). If two gas lasers are used the number of new frequencies is proportional to the product of the number of transitions of each laser, and is substantially increased by including all the relatively abundant isotopic species. As an example (A. H. M. Ross, unpublished calculation) using known laser transitions in the four most abundant isotopes of CO_2 and CO, outputs with an average spacing of the order of 300 MHz are possible near 16 μm; and by increasing the gas pressure to broaden the gain bandwidths, continuous coverage should be possible. By mixing 97 mW of CO radiation with 1.25 W of CO_2 radiation in a crystal of $CdGeAs_2$, 4 μW of power has been generated at 13 μm (Kildal and Mikkelsen, 1974). For doubling of CO_2 radiation 75 mW of CW power has been obtained with 17-W input power and using repetitively Q-switched CO_2 lasers up to 750 mW of average doubled power has been obtained and conversion efficiencies as high as 27% have been observed (Menyuk *et al.*, 1976). In addition, with 150-nsec CO_2 transversely excited atmospheric (TEA) laser pulses, 200 mJ of doubled radiation has been obtained (H. Kildal, personal communication) with an optical conversion efficiency of 10%. Dye laser outputs have been mixed in $AgGaS_2$ to obtain radiation between 4.6 and 18 μm (Hanna *et al.*, 1973; Seymour *et al.*, 1976). Noncollinear phase matching has also been used to mix infrared radiation in GaAs (Aggarwal *et al.*, 1973; Lee *et al.*, 1976). Large area beams and folded geometries are employed to provide long interaction distances.

Sum- and difference-frequency generation can also be obtained over a narrow frequency range by modulating gas laser radiation in a nonlinear crystal with a microwave source (Corcoran *et al.*, 1970, 1973; Bicknell, *et al.*, 1975; G. M. Carter and L. R. Tomasetta, personal communication). The technique has been used to measure pressure-broadened gain bandwidths for the CO_2 laser (Corcoran *et al.*, 1973). A tuning range of several gigahertz has been demonstrated at a frequency offset of 15 GHz with 30 mW of modulated power out for inputs of 50 W of CO_2 laser power and 100 W of microwave power (G. M. Carter and L. R. Tomasetta, personal communication).

Four-wave parametric mixing in alkali metal vapors using dye lasers has yielded tunable radiation in the 2–25-μm region with 0.2-cm^{-1} linewidth (Wynne *et al.*, 1974). In this case the average power was very low: 5×10^{-8} W at 2 μm and 5×10^{-11} W at 25 μm. Radiation near 8.6 μm has been obtained by efficient four wave mixing of 9.6 and 10.6 μm CO_2 radiation in Ge. With $\sim$ 1 J in each CO_2 beam, about 10 mJ of 8.6-μm radiation has been achieved (Aggarwal *et al.*, 1976).

The optical parametric oscillator (OPO) closely resembles microwave parametric oscillators and amplifiers, and its operation is similar to difference-frequency generation (Harris, 1969; Smith, 1972; Byer, 1975). The basic OPO consists of a nonlinear crystal between two wavelength-selective mirrors to form an optical cavity. A laser field at a frequency ω_p (the pump frequency) is applied to the crystal, usually through one of the end mirrors. Initially, the pump radiation mixes with photon noise in the crystal, leading to a buildup of radiation at two frequencies: the signal frequency ω_s and the idler frequency ω_i ($= \omega_p - \omega_s$), which are mutually phase matched for difference-frequency generation with the pump. If losses in the cavity are less than the gain of the buildup process, oscillation occurs. In order to change the phase-matched wavelengths (thereby tuning the oscillator), the indices of refraction of the crystal are varied with temperature, crystal rotation, or electric field. A singly resonant OPO is one in which either the signal or the idler is fed back in the optical cavity, while in a doubly resonant oscillator both the signal and idler are fed back. A schematic of a singly resonant OPO is shown in Fig. 4 (Byer *et al.*, 1975).

Optical parametric oscillators operate CW as well as pulsed. Although CW thresholds can be as low as 2.8 mW (Harris, 1969; Smith, 1972; Byer, 1975), the CW OPO is generally difficult to operate, tends to be unstable, and requires careful geometric design. A singly resonant $LiNbO_3$ oscillator, pumped with a ruby laser, has yielded peak power up to 340 kW and has had up to 45% conversion efficiency. The average power was as high as 350 mW near 2.1 μm, with a conversion efficiency of 70% when an internal oscillator with a $LiNbO_3$ crystal in a repetitively *Q*-switched Nd:YAG laser cavity was used (Ammann, 1970). $LiIO_3$ has also been used as the nonlinear material in parametric oscillators (Goldberg, 1970; Izrailenko *et al.*, 1970; Campillo and Tang, 1971); an internal Nd:YAG pumped oscillator has produced 13-kW peak power and 20-mW average power (Goldberg, 1972). A CdSe oscillator has been operated with a 2.36-μm

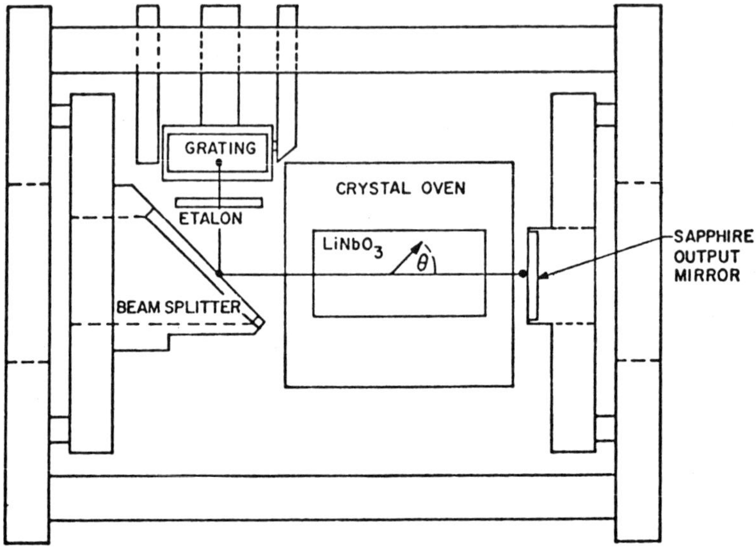

Fig. 4. Schematic of a 1.06-μm pumped, angle tuned $LiNbO_3$ singly resonant optical parametric oscillator. The pump beam enters from the left. (From Byer *et al.*, 1975.)

$Dy{:}CaF_2$ laser pump (Davidov *et al.*, 1972), with a 1.83-μm Nd:YAG pump (Herbst and Byer, 1972; Byer, 1972) and with a 2.7–3.0-μm HF laser pump (Weiss and Goldberg, 1974; Wenzel and Arnold, 1976; see also Rockwood, 1976). With an input of 80 mJ of 2.8-μm HF radiation a grating tuned output of 1 mJ and a linewidth of $0.3\,cm^{-1}$ has been achieved at 16 μm (Wenzel and Arnold, 1976; see also Rockwood, 1976). In this system tuning was obtained in the 14.1–16.4-μm range. No single OPO has been tuned over a very wide region because of difficulties with mirror coatings and materials. A $LiNbO_3$ oscillator pumped by intracavity doubled radiation from a *Q*-switched Nd:YAG laser has been developed as a commercial device (Wallace, 1970). This system, with changes of mirrors on both the oscillator and pump laser, operates anywhere between 0.55 and 3.5 μm, with peak powers of up to several kilowatts and average powers of over 100 mW. Long-term linewidths of less than $0.05\,cm^{-1}$ have been achieved. Although interferometric techniques have allowed single-mode operation in several systems, long-term stability has not been carefully measured. One technique for long-term stability is to lock the OPO wavelength to a gas absorption line. Linewidths of less than $0.001\,cm^{-1}$ have been demonstrated for an OPO operating at 2.5 μm (Pinnard and Young, 1972).*

* Chromatix Corporation, Mountain View, California has developed an OPO with a similar bandwidth.

III. Semiconductor Diode Lasers

Recombination-radiation semiconductor lasers (Nathan, 1966; Kressel, 1971; Harman, 1970; Melngailis and Mooradian, 1975) operate by stimulating emission across the gap between conduction and valence bands. Population inversion is achieved by electron injection across the band gap either with an electrical current (diode) or by optical pumping or electron-beam excitation. Infrared semiconductor laser materials in the 1–30-μm region include such binary compounds as InAs, InSb, GaSb, PbSe, PbS, PbTe, and pseudobinary alloys such as $Pb_{1-x}Sn_xTe$, $PbS_{1-x}Se_x$, $Hg_{1-x}Cd_xTe$, $In_xGa_{1-x}As$, and $GaAs_xSb_{1-x}$. Materials such as GaAs, InP, $Ga_xAl_{1-x}As$, $GaAs_xP_{1-x}$, and $GaAs_xSb_{1-x}$ can be used as tunable lasers in the visible and near infrared. I shall discuss here only the lead salt diode lasers because these devices are most useful in infrared spectroscopy. In Fig. 5 a lead salt diode laser is shown mounted in a header which provides the heat sinking and the electrical current.

Fig. 5. Photograph of a lead salt diode laser. The diode is the small rectangular crystal ($\sim 200 \times 250 \times 500$ μm) sandwiched between the bottom support stud and the upper contact bar. The support stud and the contact bar provide both electrical contact and heat sinking. (Package designed by N. DeMeo and A. R. Calawa.)

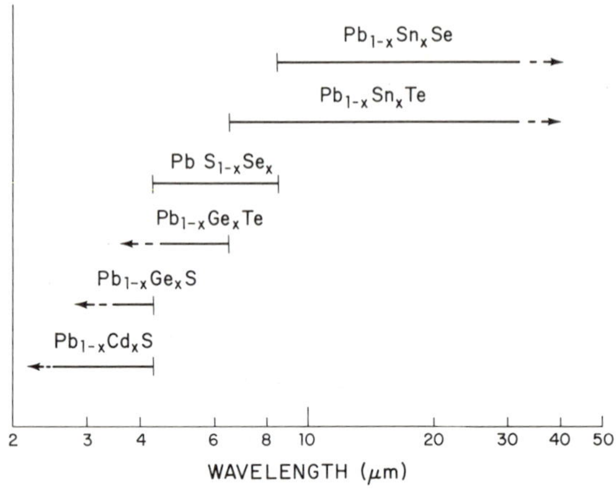

Fig. 6. Compositional tuning ranges of lead salt materials.

As shown in Fig. 6 the infrared frequency of lead salt diode lasers can be coarse tuned by adjusting the chemical composition (i.e., the factor x in $Pb_{1-x}Sn_xTe$ or $PbS_{1-x}Se_x$ crystals), which determines the energy gap of the semiconductor and thus the wavelength of the spontaneous emission. Coarse tuning of the finished lasers can be accomplished by changing the temperature, applied pressure, or magnetic field. Fine tuning can also be obtained by changing these parameters, which affects the index of refraction and thus the cavity mode frequency.

Most diode lasers emit radiation in several modes whose infrared frequency increases with increased drive current (Hinkley *et al.*, 1968). Continuous tuning of a single mode can be several reciprocal centimeters before a mode jump occurs, but is usually 0.5–1.0 cm^{-1}. The spontaneous emission from these devices tunes of the order of 300 cm^{-1} between 4° and 77°K (Harman *et al.*, 1969), and the cavity tuning rates for typical lead salt lasers are about one-third of the spontaneous rate. The continuous tuning range of a single mode is quite adequate for scanning a pressure-broadened gas absorption line in the infrared, which is typically about 0.1 cm^{-1} wide.

The magnetic-field dependence of wavelength in diode lasers has been extensively investigated (Melngailis and Mooradian, 1975; Butler and Calawa, 1966; Calawa *et al.*, 1969). Because of the large mass anisotropy in the lead salts, the magnetic energy levels depend on the orientation of the applied magnetic field with respect to the crystallographic axes. Both the conduction and valence bands have their energies quantized into Landau levels, each of which is split into two spin energy states. Usually, electrons injected into the conduction band thermalize to the lowest available energy state, while holes injected into the valence band thermalize to the highest available energy state before recombining. Fine tuning of the axial modes with magnetic field occurs in a similar fashion to that of temperature tuning.

The tuning rate within a mode (and for different modes) can vary with magnetic field; and for a $PbS_{0.82}Se_{0.18}$ diode laser has been measured and found to vary from 0.4 to 2.0 MHz G^{-1} (Nill *et al.*, 1973).

Application of hydrostatic pressure to a semiconductor laser can provide a very broad tuning range for a single device (Besson *et al.*, 1956, 1968). A PbSe diode laser at 77°K has been tuned from 7.5 to 22 μm using hydrostatic pressure up to 14 kbar. More recently, the stability of the pressure has been controlled sufficiently to perform Doppler-limited spectroscopy using GaAs (Pine *et al.*, 1973) and PbS_xSe_{1-x} (Pine, personal communication) lasers. Uniaxial pressure tuning of diode lasers can avoid the limitations of low-temperature operation; however, in practice it is difficult to keep from crushing the devices. Uniaxial pressure tuning rates are about one-third the hydrostatic rates. The recent development of the infrared diodes that operate continuously at 77°K or higher (Groves *et al.*, 1974) will greatly facilitate the usefulness of hydrostatic pressure tuning. Because of the broad-band wavelength coverage using hydrostatic pressure tuning, only the binary semiconductor compounds and one or two alloy semiconductors would be necessary to cover the wavelength range from 2 to 35 μm.

An increase in device yield as well as internal quantum efficiency has partly resulted from improvements in the size and quality of vapor-grown crystals (Harman *et al.*, 1974). In the device technology area the dc current carrying capability has been greatly increased by new low resistance contacting and heatsinking techniques. The external efficiency has been increased by improving the optical quality of the end mirrors (Walpole *et al.*, 1973) by use of the stripe geometry to suppress parasitic modes (Ralston *et al.*, 1973) and by the application of optical coatings to the end mirrors. Lead sulfide stripe geometry lasers have given over 300-mW CW multimode with about one-fifth of the power available for single-mode operation. Junctions have been made by self-diffusion (Walpole *et al.*, 1973a) as well as by antimony (Antcliffe and Parker, 1973) and cadmium (Lo, 1976) diffusion.

Lead salt heterostructure lasers have been fabricated (Groves *et al.*, 1974; Walpole *et al.*, 1973b) in an effort to reduce the high-temperature threshold and to achieve CW operation above 77°K. As previously demonstrated in GaAs–$Ga_{1-x}Al_xAs$ heterostructure lasers, the confinement of both the injected carriers and of the light wave to an active region that is bounded by epitaxially grown regions of wider energy gap results in a substantial decrease in laser threshold. As a first step in developing such structures in the lead salts a single heterojunction was made by evaporating a layer of *n*-type PbTe on a *p*-type $Pb_{1-x}Sn_xTe$ substrate (Walpole *et al.*, 1973b). While the low-temperature threshold of the heterostructure laser is comparable to the lowest observed in homojunction devices, at 77°K the threshold is considerably lower than previously observed in homojunctions. Even though the preliminary devices were not constructed with adequate heat sinking (the stripe was not in contact with the heat sink), CW operation was observed up to 65°K.

Double heterostructure stripe-geometry lasers have been fabricated by means of liquid phase epitaxy (Groves *et al.*, 1974). An active layer of *n*-type $Pb_{0.88}Sn_{0.12}Te$ was grown on a Te-doped PbTe substrate in stripe openings of a MgF_2 coating on the PbTe. This was followed by the growth of a confining layer of *n*-type PbTe. The CW threshold was 1.6×10^3 A cm^{-2} at 12°K and 4.2×10^3 A cm^{-2} at 77°K. A cw power of 10 mW in four modes was measured at 12°K and single mode operation was observed at 77°K with 1.2-mW power. The laser tuned from 10.5 μm at 12°K to 8.2 μm at 80°K, a range of 280 cm^{-1}.

Recently, double-heterostructure $Pb_{1-x}Sn_xTe$ lasers with active regions of $Pb_{0.782}Sn_{0.218}Te$ have been grown by molecular-beam epitaxy and operate CW up to heat-sink temperatures of 114°K (Walpole *et al.*, 1976). Temperature tuning of the emission from 15.9 to 8.54 μm wavelength has been obtained, with emission at 77°K near 11.5 μm. Figure 7 shows output wave-

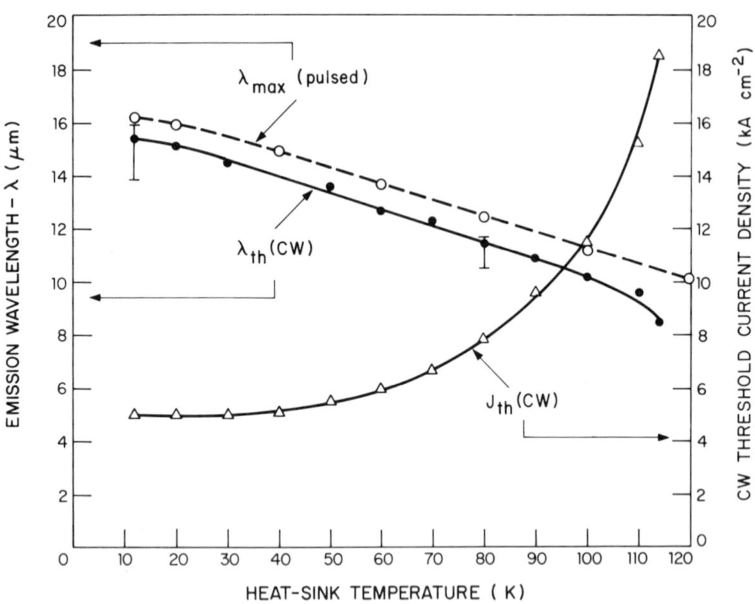

Fig. 7. The solid curve through the experimental points (solid dots) gives the wavelength of emission of the threshold mode versus heat-sink temperature. The range of emission at heat-sink temperatures of 12 and 80°K is indicated by the vertical lines for current densities of several kA cm^{-2} above threshold. The dashed curve through the experimental points (open circles) gives the longest wavelength mode seen under low-duty-cycle pulsed operation as a function of heat-sink temperature. The curve through the triangular data points gives the CW threshold current density as a function of heat-sink temperature. (From Walpole *et al.*, 1976.)

length and threshold current versus temperature for this device. The current–voltage characteristics showed an abrupt change in slope at threshold, indicating high incremental internal quantum efficiency.

IV. Raman Lasers

The spin-flip Raman (SFR) laser (Patel and Shaw, 1971; Mooradian *et al.*, 1971; Eng *et al.*, 1974; Stattler *et al.*, 1974) is a device that uses a fixed-frequency laser (CO_2, CO, or HF gas laser) to pump a semiconductor crystal in a magnetic field. The pump-laser photons lose energy when they interact with an electron in the crystal and flip its spin. The down-shifted Raman photon is separated in energy from the pump photon by the change in electron spin energy $g\beta H$, where g is the gyromagnetic ratio of the conduction electron, β the Bohr magneton, and H the magnetic field strength. Consequently, the Raman photon frequency depends on magnetic field. At sufficiently high pump power, stimulated emission of Raman photons can exceed losses, and exponential gain and oscillation occurs.

In the most widely studied SFR laser, n-type InSb is the semiconductor crystal. The laser has been operated in the pulsed mode (Patel and Shaw, 1971) using CO_2 or CO laser pumps, and in the CW mode (Mooradian *et al.*, 1971) with a CO laser pump. Under CW conditions, short-term stability relative to the CO pump laser of better than 1 kHz has been achieved. (Patel, 1972). A hybrid frequency and phase-lock closed-loop system was constructed (Brueck and Mooradian, 1974) to maintain the output frequency of the SFR laser at a variable frequency offset from that of the CO local oscillator laser. The error signal derived by heterodyning the SFR laser output and stable CO laser output together with the output from a stable microwave oscillator, in a HgCdTe photodetector, has been used to drive small modulation coils around the InSb sample. A schematic of this system is shown in Fig. 8.

The resonance that exists when the pump laser is at a frequency near the bandgap of the semiconductor allows CW operation with low pump threshold power (Mooradian *et al.*, 1971). Less than 50 mW of TEM_{00} pump-laser power in an estimated 50–100-μm focal diameter in the crystal has been found sufficient to achieve threshold for the case in which incident pump and Raman beams were collinear, and a maximum output power in excess of 1-W CW was obtained. In this case, over 75% of the pump energy was converted to the first down-shifted Raman mode. At high conversion efficiencies, the beam divergence of the SFR laser is less than 40 mrad, consistent with a TEM_{00} mode divergence from a 250-μm diameter laser

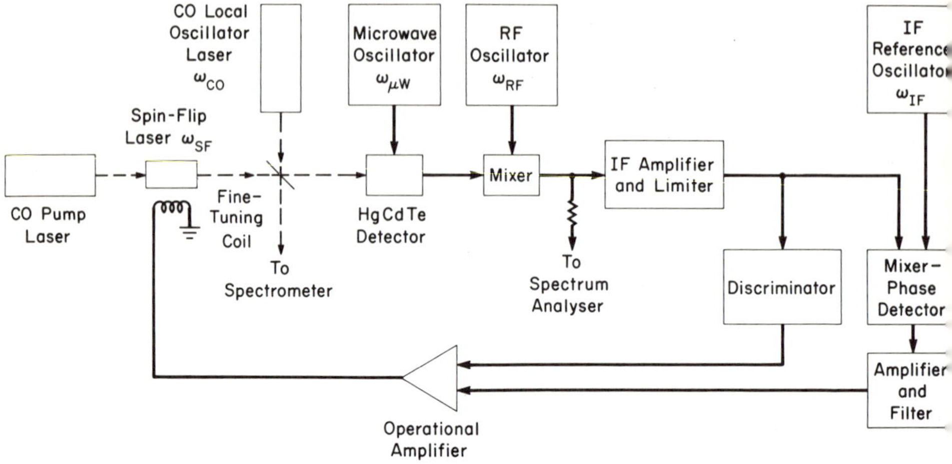

Fig. 8. Hybrid frequency and phase-lock system. (From Brueck and Mooradian, 1974.)

cavity. Operation in single axial and single transverse modes has been observed, as well as mode hopping during magnetic field tuning. About 40% of the overall tuning range is covered, with continuous tuning in each mode of approximately $0.3\,cm^{-1}$. Mode pulling caused by changing the wavelength of the emission peak results in fine tuning, in contrast to the diode laser, in which a change in re ractive index directly tunes the laser wavelength.

As part of an effort to improve pulsed spin-flip laser operation, measurements (Broeck and Mooradian, 1976) have been made of the electron spin population relaxation time in InSb. Spin-flip gain is reduced by free-carrier heating effects, and the recovery time for this process has been measured. The pump laser system for these observations consisted of a CO_2 TEA laser with a CW gain section to control the temporal and spatial output structure. An external CdTe modulator system was constructed that provided a train of pump laser radiation pulses whose width and spacing could be varied. The relaxation times were measured by observing the temporal behavior of the spin-flip laser output as a function of pump pulse train parameters. For a sample at $T \sim 20°K$ having a conduction electron concentration of $1.2 \times 10^{16}\,cm^{-3}$ the spin population relaxation time was found to be 60 ± 20 nsec. A comparable time was found for recovery from free-carrier heating.

Stimulated SFR scattering in InAs has been obtained (Eng *et al.*, 1974) using an HF–TEA laser ($2 \rightarrow 1$ band, P(9) line at $3385.34\,cm^{-1}$) to pump near the band-gap resonance. Tunable emission was observed from 3347 to $3332\,cm^{-1}$. The output power in the first Stokes component increased monotonically with field, with conversion efficiencies of more than 10% measured. Stimulated SFR scattering near $10\,\mu m$ has also been reported using CO_2 TEA lasers (Stattler *et al.*, 1974) and Q-switched lasers (Kruse, 1976) to resonantly pump an n-type crystal of $Hg_xCd_{1-x}Te$.

The use of stimulated Raman scattering to downshift tunable dye laser radiation into the infrared has been carried out using high-pressure hydrogen (Schmidt and Appt, 1972, 1974) and atomic vapors (Wynne and Sorokin, 1975; Carlsten and Dunne, 1975; Cotter *et al.*, 1975, 1976). In hydrogen, with 12-kW of dye laser pump power, third Stokes emission was observed at 2.1 μm (Schmidt and Appt, 1972, 1974). Output powers of 7 kW have been observed in the 5.67–8.65-μm range using cesium and 0.1–1.0 MW of dye laser pump power (Cotter *et al.*, 1976). Stimulated Raman scattering in hydrogen of tunable neodymium glass laser radiation has given 0.8 J of output in the 8.33–9.1-μm range (Grasiuk *et al.*, 1976; Grasiuk and Zubarev, 1976). Four-wave mixing techniques can be used to further extend the utility of Raman scattering in hydrogen (Ducuing, 1976; Byer, 1976).

V. High-Pressure Gas Lasers

Molecular gas lasers, such as CO_2, have most frequently been operated at pressures of around 10 torr. Consequently, the gain bandwidths are essentially Doppler limited to approximately 50 MHz. It is desirable to increase the range of continuous tuning, and for this higher pressures are needed to broaden the gain bandwidth. However, high-pressure discharges in a long laser tube are difficult to maintain, since in addition to the increased bandwidth the deactivation rate is higher, leading to a higher pump threshold. A number of improvements in laser technology have allowed the development of high-pressure gas lasers to proceed. These include: (1) transversely-excited atmospheric (TEA) configuration lasers with photoionization (Levine and Javan, 1973) and photopreionization (Alcock *et al.*, 1973); (2) optically pumped lasers (Chang and Wood, 1973, 1974; Kildal and Deutsch, 1976; Chang *et al.*, 1976); (3) electron-beam-pumped lasers (Bagratashuili *et al.*, 1973, 1976; Harris *et al.*, 1974; O'Neill and Whitney, 1975, 1976); and (4) capillary lasers (Abrams, 1974), including pulser-sustainer devices (Smith, 1974). The first three of these devices have operated in the 10-atm range, where continuous tuning between rotational transitions is possible while the capillary system has operated to several atmospheres pressure. A 15-atm electron-beam-pumped CO_2 laser emitting 0.1 J, 40-nsec pulses has been reported (O'Neill and Whitney, 1975), and using an etalon $<0.2\,cm^{-1}$ wide radiation was obtained and tuned continuously over $2.3\,cm^{-1}$. Tunable high pressure operation has also been obtained by this method in N_2O and CS_2 (O'Neill and Whitney, 1976). Continuous tuning over $5\,cm^{-1}$ has

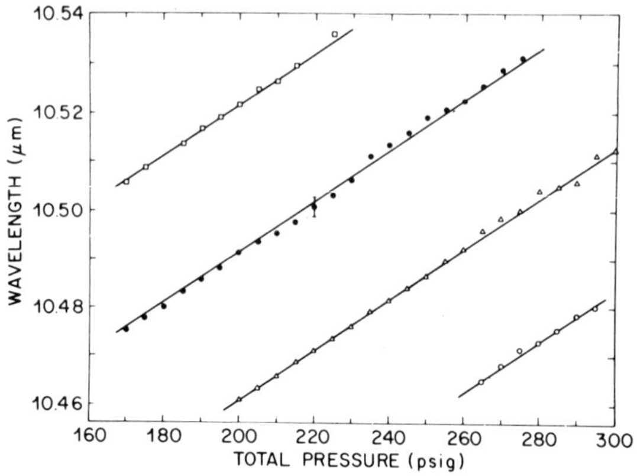

Fig. 9. Experimentally measured output wavelength from a high-pressure N_2O/CO_2 transfer laser as a function of laser operating pressure. Each line is a longitudinal mode. (From Chang *et al.*, 1976.)

been obtained recently in an optically pumped high-pressure N_2O/CO_2 transfer laser (Chang *et al.*, 1976). The laser was tuned by changing pressure and a resolution of $0.014\,cm^{-1}$ obtained. Figure 9 shows output wavelength versus cell pressure for this system. While these lasers have great potential for providing tunable infrared radiation, some effort remains in the engineering of them into practical devices. High-pressure gas lasers should have higher peak and average power than any other currently conceivable tunable infrared source.

Acknowledgments

This work was sponsored by the Department of the Air Force.

References

Abrams, R. L. (1974). *Appl. Phys. Lett.* **24**, 304.

Aggarwal, R. L., Lax, B., and Favrot, G. (1973). *Appl. Phys. Lett.* **22**, 329.

Aggarwal, R. L., Lee, N., and Lax, B. (1976). *In* "Tunable Lasers and Applications" (A. Mooradian, T. Jaeger, and P. Stokseth, eds.), p.96. Springer-Verlag, Berlin and New York.

Alcock, A. J., Leopold, K., and Richardson, M. C. (1973). *Appl. Phys. Lett.* **23**, 562.

Ammann, E. O., Yarborough, J. M., Oshman, M. K., and Montgomery, P. C. (1970). *Appl. Phys. Lett.* **16**, 309.

Antcliffe, G. A., and Parker, S. G. (1973). *J. Appl. Phys* **44**, 4145.

Bagratashuili, V. N., Knyazev, I. N., Kudryatsev, Yu. A., and Letokhov, V. S. (1973). *JETP Lett.* **18**, 62.

Bagratashvili, V. N., Knyazev, I. N., Letokhov, V. S., and Lobko, V. V. (1976). *Sov. J. Quantum Electron.* **5**, 541.

Besson, J. M., Butler, J. F., Calawa, A. R., Paul W., and Rediker, R. H. (1956). *Appl. Phys. Lett.* **7**, 206.

Besson, J. M., Paul, W., and Calawa, A. R. (1968). *Phys. Rev.* **173**, 699.

Bethea, C. G. (1973). *Appl. Opt.* **12**, 1104.

Bhar, G. C., Hanna, D. C., Luther-Davies, B., and Smith, R. C. (1972). *Opt. Commun.* **6**, 323.

Bicknell, W. E., Tomasetta, L. R., and Kingston, R. H. (1975). *IEEE J. Quantum Electron.* **11**, 308.

Boyd, G. D., Buehler, E., Storz F. G., and Wernick, J. H., (1972a). *IEEE J. Quantum Electron.* **8**, 419.

Boyd, G. D., Kasper, H. M., McFee, J. H., and Storz, F. G. (1972b). *IEEE J. Quantum Electron.* **8**, 900.

Bridges, T. J., Nguyen, V. T., Burkhardt, E. G., and Patel, C. K. N. (1975). *Appl. Phys. Lett.* **27**, 600.

Brueck, S. R. J., and Mooradian, A. (1974). *IEEE J. Quantum Electron.* **10**, 634.

Brueck, S. R. J., and Mooradian, A. (1976). *Conf. Digest, Int. Quantum Electron. Conf., 9th, Amsterdam, The Netherlands*, p. 15.

Butler, J. F., and Calawa, A. R. (1966). *In* "Physics of Quantum Electronics" (P. L. Kelley, B. Lax, and P. E. Tannenwald, eds.), p. 458. McGraw-Hill, New York.

Byer, R. L. (1972). *Digest Tech. Papers, Int. Quantum Electron. Conf., 7th, Montreal, Canada*, p. 57.

Byer, R. L. (1976). *In* "Tunable Lasers and Applications" (A. Mooradian, T. Jaeger, and P. Stokseth, eds.), p. 70. Springer-Verlag, Berlin and New York.

Byer, R. L. (1975). *In* "Quantum Electronics" (H. Rabin and C. L. Tang, eds.), Vol. 1B, p. 587. Academic Press, New York.

Byer, R. L., Kildal, H., and Feigelson, R. S. (1971). *Appl. Phys. Lett.* **19**, 237.

Byer, R. L., Herbst, R. L., and Fleming, R. N. (1975). *In* "Laser Spectroscopy" (S. Haroche, J. C. Pebay-Peyroula, T. W. Hänsch, and S. E. Harris, eds.), p. 207. Springer-Verlag, Berlin and New York.

Calawa, A. R., Dimmock, J. O., Harman, T. C., and Melngailis, I. (1969). *Phys. Rev. Lett.* **23**, 7.

Campillo, A. J., and Tang, C. L. (1971). *Appl. Phys. Lett.* **12**, 376.

Carlsten, J. L., and Dunn, P. C. (1975). *Opt. Commun.* **14**, 8.

Chang, T. Y., and Wood, O. R. (1973). *Appl. Phys. Lett.* **23**, 524.

Chang, T. Y., and Wood, O. R. (1974). *Appl. Phys. Lett.* **24**, 182.

Chang, T. Y., McGee, J. D., and Wood, O. R. (1976). *Opt. Commun.* **18**, 279.

Chemla, D. S., Kupeck, P. J., Robertson, D. S., and Smith, R. C. (1971). *Opt. Commun.* **3**, 29.

Corcoran, V. J., Cupp, R. E., Gallagher, J. J., Smith, W. T. (1970). *Appl. Phys. Lett.* **16**, 316.

Corcoran, V. J., Martin, J. M., and Smith, W. T. (1973). *Appl. Phys. Lett.* **22**, 517.

Cotter, D., Hanna, D. C., Karkkainen, P. A., and Wyatt, R. (1975). *Opt. Commun.* **15**, 143.

Cotter, D., Hanna, D. C., and Wyatt, R. (1976). *Opt. Commun.* **16**, 256.

Davidov, A. A., Kulevskii, L. A., Prokhorov, A. M., Savelev, A. D., and Smirnov, V. V. (1972). *JETP Lett.* **15**, 513.

Decker, C. D., and Tittel, F. K. (1973). *Opt. Commun.* **8**, 244.

Dewey, C. F., Jr., and Hocker, L. O. (1971). *Appl. Phys. Lett.* **18**, 58.

Ducuing, J., Frey, R., and Pradère, F. (1976). *In* "Tunable Lasers and Applications" (A. Mooradian, T. Jaeger, and P. Stokseth, eds.), p. 81. Springer-Verlag, Berlin and New York.

Eng, R. S., Mooradian, A., and Fetterman, H. R. (1974). *Appl. Phys. Lett.* **25**, 453.

Goldberg, L. S. (1970). *Appl. Phys. Lett.* **17**, 489.

Goldberg, L. S. (1972). *Digest Tech. Papers, Int. Quantum Electron. Conf., 7th, Montreal, Canada*, p. 55.

Goldberg, L. S. (1975). *Appl. Opt.* **14**, 653.

Grasiuk, A. Z., and Zubarev, I. G. (1976). *In* "Tunable Lasers and Applications" (A. Mooradian, T. Jaeger, and P. Stokseth, eds.), p. 88. Springer-Verlag, Berlin and New York.
Grasiuk, A. Z., Zubarev, I. G., Kotov, A. V., Mikhailov, S. I., and Smirnov, V. G. (1976). *Sov. J. Quantum Electron.* **6**, 568.
Groves, S. H., Nill, K. W., and Strauss, A. J. (1974). *Appl. Phys. Lett.* **25**, 331.
Hanna, D. C., Smith, R. C., and Stanley, C. R. (1971). *Opt. Commun.* **4**, 300.
Hanna, C., Rampal, V. V., and Smith, R. C. (1973). *Opt. Commun.* **8**, 151.
Hanna, D. C., Luther-Davies, B., Smith, R. C., and Wyatt, R. (1974). *Appl. Phys. Lett.* **25**, 142.
Harman, T. C. (1970). *In* "The Physics of Semimetals and Narrow-Gap Semiconductors," p. 363. Pergamon, Oxford.
Harman, T. C., and McVittie, J. P. (1974). *J. Electron. Mater.* **3**, 843.
Harman, T. C., Calawa, A. R., Melngailis, I., and Dimmock, J. O. (1969). *Appl. Phys. Lett.* **14**, 333.
Harris, N. W., O'Neill, F., and Whitney, W. T. (1974). *Appl. Phys. Lett.* **25**, 148.
Harris, S. E. (1969). *Proc. IEEE* **57**, 2096.
Herbst, R. L., and Byer, R. L. (1972). *Appl. Phys. Lett.* **21**, 189.
Hinkley, E. D., Harman, T. C., and Freed, C. (1968). *Appl. Phys. Lett.* **13**, 49.
Izrailenko, A. I., Kovrigin, A. I., and Nikles, P. V. (1970). *JETP Lett.* **12**, 331.
Kildal, H., and Deutsch, T. F. (1976). *Conf. Digest. Int. Quantum Electron. Conf. 9th, Amsterdam*, p. 124.
Kildal, H., and Mikkelsen, J. C. (1973). *Opt. Commun.* **9**, 315.
Kildal, H., and Mikkelsen, J. C. (1974). *Opt. Commun.* **10**, 306.
Kildal, H., Iseler, G. W., Menyuk, N., and Mikkelsen, J. C. (1976). *Tech. Digest, OSA Topical Meeting Opt. Phenomena Infrared Mater., Annapolis, Maryland.*
Kressel, H. (1971). *In* "Lasers" (A. K. Levine and A. J. DeMaria, eds.), Vol. 3, p. 1. Dekker, New York.
Kruse, P. W. (1976). *Appl. Phys. Lett.* **28**, 90.
Lee, N., Lax, B., and Aggarwal, R. L. (1976). *Conf. Digest, Int. Quantum Electron. Conf., 9th, Amsterdam, The Netherlands*, p. 50.
Levine, J. S., and Javan, A. (1973). *Appl. Phys. Lett.* **22**, 55.
Lo, W. (1976). *Appl. Phys. Lett.* **28**, 154.
Melngailis, I., and Mooradian, A. (1975). *In* "Laser Applications to Optics and Spectroscopy" (S. F. Jacobs, M. Sargent, J. F. Scott, and M. O. Scully, eds.), p. 1. Addison-Wesley, Reading, Massachusetts.
Meltzer, D. W., and Goldberg, L. S. (1972). *Opt. Commun.* **5**, 209.
Menyuk, N., Iseler, G. W., and Mooradian, A. (1976). *Appl. Phys. Lett.* **29**, 422.
Mooradian, A. (1974). *In* "Laser Spectroscopy" (R. G. Brewer and A. Mooradian, eds.), p. 223. Plenum Press, New York.
Mooradian, A., Brueck, S. R. J., and Blum, F. A. (1971). *Appl. Phys. Lett.* **17**, 481.
Nathan, M. I. (1966). *Proc. IEEE* **54**, 1276.
Nill, K. W., Blum, F. A., Calawa, A. R., and Harman, T. C. (1973). *J. Nonmetals* **1**, 211.
O'Neill, F., and Whitney, W. T. (1975). *Appl. Phys. Lett.* **26** 454.
O'Neill, F., and Whitney, W. T. (1976). *Appl. Phys. Lett.* **28**, 539.
Patel, C. K. N. (1972). *Phys. Rev. Lett.* **28**, 649.
Patel, C. K. N., and Shaw, E. D. (1971). *Phys. Rev. B* **3**, 1279.
Pike, H. A., Hinkley, E. D., and Calawa, A. R. (1972). Optics Research Report, Lincoln Lab., MIT (1972:1), p. 79, DDC AD-754939.
Pine, A. S. (1974). *J. Opt. Soc. Am.* **64**, 1683.
Pine, A. S. (1976). *J. Opt. Soc. Am.* **66**, 97.
Pine, A. S., Glassbrenner, C. J., and Kafalas, J. A. (1973) *IEEE J. Quantum Electron.* **9**, 800.
Pinnard, J., and Young, J. F. (1972). *Opt. Commun.* **4**, 425.

Ralston, R. W., Melngailis, I., Lindley, W. T., and Calawa, A. R. (1973). *IEEE J. Quantum Electron.* **9**, 350.

Schmidt, W., and Appt, W. (1974). Post Deadline Paper R6, *Int. Quantum Electron. Conf., 8th, San Francisco, California.*

Schmidt, W., and Appt, W. (1972). *Z. Naturforsch.* **27a**, 1373.

Seymour, R. J., Zernike, F., and Sam, C. L. (1976). *Conf. Digest, Int. Quantum Electron. Conf., 9th, Amsterdam, The Netherlands*, p. 49.

Smith, R. G. (1972). *In* "Laser Handbook" (F. T. Arrechi and E. O. Schulz-DuBois, eds.), p. 837. North-Holland Publ., Amsterdam.

Smith, P. W. (1974). *In* "Laser Spectroscopy" (R. G. Brewer and A. Mooradian, eds.), p. 247. Plenum Press, New York.

Stattler, J. P., Weber, B. A., and Nemarich, J. R. (1974). *Appl. Phys. Lett.* **25**, 491.

Wallace, R. W. (1970). *Appl. Phys. Lett.* **17**, 497.

Walpole, J. N., Calawa, A. R., Ralston, R. W., and Harman, T. C. (1973a). *J. Appl. Phys.* **44**, 2905.

Walpole, J. N., Calawa, A. R., Ralston, R. W., Harman, T. C., and McVittie, J. P. (1973b). *Appl. Phys. Lett.* **23**, 620.

Walpole, J. N., Calawa, A. R., Harman, T. C., and Groves, S. H. (1976). *Appl. Phys. Lett.* **28**, 553.

Weiss, J. A., and Goldberg, L. S. (1974). *Appl. Phys. Lett.* **24**, 389.

Wenzel, R. G., and Arnold, G. P. (1976); *Appl. Opt.* **15**, 1322. See also S. D. Rockwood, S. D. (1976). *In* "Tunable Lasers and Applications" (A. Mooradian, T. Jaeger, and P. Stokseth, eds.), p. 140. Springer-Verlag, Berlin and New York.

Wynne, J. J., and Sorokin, P. P. (1975). *J. Phys. B* **8**, L37.

Wynne, J. J., Sorokin, P. P., and Lankard, J. R. (1974). "Laser Spectroscopy" (R. G. Brewer and A. Mooradian, eds.), p. 103. Plenum Press, New York.

6

Infrared Optoacoustic Spectroscopy and Detection

P. C. Claspy

Department of Electrical Engineering and Applied Physics
Case Western Reserve University
Cleveland, Ohio

I. Introduction

Systems using the optoacoustic effect in molecular gases have been widely used in gas analysis and vapor detection since the original descriptions of this application by Veingerov (1938) and by Luft (1943). These nondispersive infrared (NDIR) systems, which employ a blackbody source, a nondispersive optical path, and an optoacoustic detector containing the gas component to be analyzed, are discussed extensively by Hill and Powell (1968) and are not considered in detail here. Their basic concept is that a gas-filled optoacoustic detector is a wavelength-selective device and can therefore measure attenuation by the medium between source and detector of those wavelengths absorbed by the detector. The system thus performs a measurement of incident and transmitted intensity, from which absorption coefficient or absorber concentration can be calculated. These systems are susceptible to spurious effects arising from beam attenuation from scattering by dust or dirt. In addition, since a wide-band source is used, elaborate but only modestly successful measures must be taken, because of overlapping absorption bands, to insure correct identification of the gas species to be detected (Dailey, 1967).

Since the optoacoustic signal is proportional to the amount of power absorbed, and to the concentration of the absorbing species in the detector, direct detection of a vapor by insertion of the gas sample into the detector and using monochromatic excitation provides improvements in both sensitivity and accuracy over the nondispersive system. As an example of this, consider the detection of CO_2 using an 800°K blackbody source. With such a source it is possible to obtain an average power of $5\,\mu$W in a 0.04-μm bandwidth around 4.283 μm, which corresponds to a fundamental absorption of CO_2. Dewey (1974) calculates that using such a source with a resonant optoacoustic cell a sensitivity of 0.14 ppm of CO_2 can be obtained. Hill and Powell (1968) report a maximum sensitivity for a NDIR CO_2 detector to be at best 0.5 ppm, although it is likely that the spurious effects referred to previously would increase this to a few parts per million. Thus, the direct detection optoacoustic device provides a significant improvement over commercial NDIR systems even when a blackbody source is used.

The high power available from lasers makes possible further significant increases in the sensitivity of optoacoustic detection while the narrow bandwidth of such sources gives improved selectivity over blackbody sources. The use of a laser as the exciting source was first reported by Kerr and Atwood (1968) and has subsequently been described by numerous other authors (Kreuzer, 1971; Kreuzer and Patel, 1971; Kreuzer *et al.*, 1972; Patel *et al.*, 1974; Dewey *et al.*, 1973; Goldan and Goto, 1974; Claspy *et al.*, 1976). Implicit in the discussion of the advantages of laser sources is the expectation that the required wavelengths for a particular spectroscopic study are available. While there are many tunable infrared lasers described in the literature, and progress in this area has been rapid, only a few of these are practically available as sources for optoacoustic spectroscopy.

In the remaining sections of this chapter design considerations for optoacoustic vapor detection systems, vapor detection, identification and concentration measurement, and the application of optoacoustic techniques to the measurement of molecular relaxation rates are discussed. While applications using laser excitation is emphasized, consideration is also given to systems using blackbody sources.

II. Vapor Detection-Identification and Quantitative Measurement

A. Introduction

The characteristics of high sensitivity and linearity of response with concentration make the laser-excited optoacoustic detection system a useful tool for detection and monitoring of atmospheric pollutants and in conducting measurements of concentration of infrared-absorbing molecules in the atmosphere. In addition these systems provide a capability for detection of the presence of undesirable low-vapor-pressure materials such as illegal explosives and drugs. In each of these applications the presence and/or concentration of the vapor being investigated must be determined in the atmosphere, which typically contains many other infrared-absorbing species. Thus, in designing a system, consideration must be given both to the absorption spectrum of the species to be studied and to the absorption spectrum of any anticipated interfering species.

B. General System Configuration

A typical apparatus used for laser excited optoacoustic detection is shown schematically in Fig. 1. The laser is a CO or CO_2 laser, the output of which is modulated at a frequency appropriate for the optoacoustic detector. The detector, which may be an acoustic resonator, is discussed extensively by Dewey (Chapter 3, this volume) and is not described in detail here. The reference detector is used to monitor exciting optical power. Since the laser output power is a function of the wavelength of

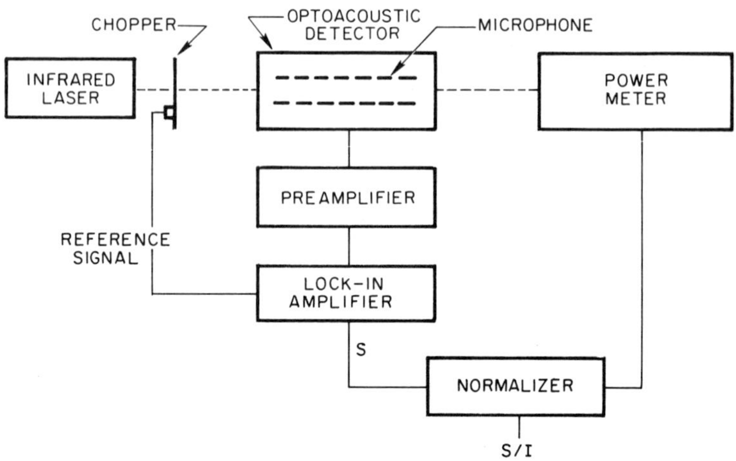

Fig. 1. Schematic diagram of a typical optoacoustic system.

operation, as well as of operating conditions, the reference detector permits normalization of the optoacoustic signals, thereby eliminating effects of laser power variations. Phase-sensitive detection of the optoacoustic signal provides significant noise reduction and resultant increase in system detectivity.

The transducer that is typically used in an optoacoustic detector to convert energy from mechanical to electrical form is a condenser microphone of either the electret type or the externally biased type. Such a microphone consists of a thin metal diaphragm or metallized plastic dielectric and a rigid conducting back plate. If a charge is applied to the capacitor by an external dc supply (electrically biased microphone), capacitance modulation caused by sound–pressure-variation-induced changes in plate separation will produce a current flow between the plates. The signal output from the microphone therefore depends on the microphone capacitance, the magnitude of the pressure induced changes in capacitance, and the magnitude of the applied voltage. As shown below, the signal is inversely proportional to the total circuit capacitance, so it is essential that the microphone capacitance, the amplifier input capacitance, and any stray circuit capacitance be kept to a minimum.

An equivalent circuit diagram of a condenser microphone is shown in Fig. 2 (Microphones and Microphone Preamplifiers, 1975). The output voltage is

$$V_0(t) = \frac{\Delta C(t)}{C_t} E_0 \frac{j\omega R C_t}{1 + j\omega R C}, \qquad (1)$$

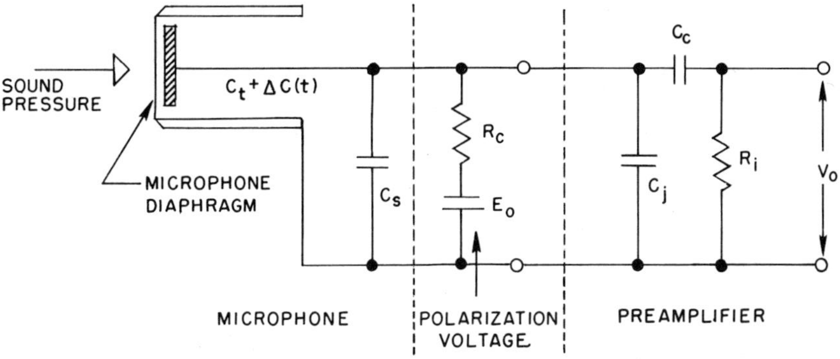

Fig. 2. Equivalent circuit diagram of a condenser microphone and preamplifier. (From Microphones and Microphone Preamplifiers, 1975.)

where E_0 is the polarization voltage, $\Delta C(t)$ the variation in capacitance resulting from sound pressure; $C = C_t + C_s + C_i$ and $R = R_i R_c/(R_i + R_c)$, where C_t is the microphone capacitance, C_s the stray capacitance (including cable capacitance), C_i the preamplifier input capacitance, C_c the coupling capacitance ($C_c \gg C$), R_c the charging resistance, and R_i the preamplifier input resistance. The sensitivity of the microphone is

$$S = \frac{V_0(t)}{\Delta p(t)} = \frac{\Delta C(t)}{C_t \, \Delta p(t)} E_0 \frac{j\omega R C_t}{1 + j\omega RC}, \tag{2}$$

where $\Delta p(t)$ is the sound pressure variation with time. Two basic cases can be considered in the analysis of this frequency response.

If $\omega RC \gg 1$, then

$$S = S_0 E_0 [C_t/(C_t + C_s + C_i)], \tag{3}$$

where $S_0 = \Delta C(t)/C_t \, \Delta p(t)$ is the open circuit sensitivity. Thus in the high-frequency limit the useful sensitivity is reduced by stray circuit capacitance and by the input capacitance of the preamplifier. For this reason the preamplifier is usually located in close proximity to the microphone.

If $\omega RC \ll 1$, then

$$S = S_0 E_0 j\omega R C_t, \tag{4}$$

and the sensitivity is frequency dependent. For this reason the preamplifier input resistance and the charging resistor must be very large ($> 5 \times 10^8 \, \Omega$). High input impedances are readily available in FET preamplifiers.

Once the basic system design is completed, the wavelengths to be used in the measurement must be selected. These are different for each vapor to be detected and depend on the infrared absorption spectra of the vapor to be studied and of any other infrared-absorbing species that may be present.

C. *Wavelength Selection*

Nearly all gases absorb in the near infrared (~2–15 μm), the most notable exceptions being the homonuclear diatomic molecules. This entire range, however, is not practically useful for laser-excited optoacoustic detection since only the CO and CO_2 lasers are conveniently available sources. While sources such as tunable infrared diode lasers and spin-flip Raman lasers have been used, practical considerations typically confine measurements to the ~5–7-μm range of the CW CO laser and 9.2–11.5 μm of the CW CO_2 laser. Tables of wavelengths available from CO_2 and CO lasers are given in the Appendix to this chapter.

Although neither the CO laser nor the CO_2 laser is continuously tunable (line separations range from ~2–4 cm^{-1}), frequent coincidences or near coincidences occur between laser emission lines and vapor absorption lines, enabling efficient optoacoustic detection. This discrete tunability is in fact a distinct advantage in detection systems, especially those directed at detection of small molecules. The advantage arises from the high degree of frequency stability and reproducibility that is easily achieved with a gas laser (Freed, 1970, 1971). This characteristic insures the availability of a source which provides a sequence of well-defined, accurately reproducible wavelengths. Therefore, once measurement wavelengths have been selected, consistent, reliable analysis is possible.

The selection of the specific spectral region in which to perform analysis or detection is based primarily on a knowledge of the infrared spectrum of the molecule to be studied. As Gelbwachs (1974) has observed, the structures of the absorption spectra of light and heavy molecules are typically quite different. For light molecules such as the normal atmospheric pollutants, spectral details are frequently resolved even at pressures as high as 1 atm, whereas heavy molecules such as drugs and explosives exhibit unresolved continuum spectra over several wavenumbers in this pressure range. Because of this difference in the spectra of the two types of molecules, the detection strategy is somewhat different in the two cases. In the case of light-molecule detection it is desirable to use laser wavelengths that coincide with some of the absorption maxima in the resolved spectrum, while in the case of heavy-molecule detection the use of a sequence of laser wavelengths which span the unresolved continuum is indicated.

Detection and identification methods for small-molecule gases and for large-molecule gases are discussed in the next two sections.

D. *Detection and Identification of Small-Molecule Gases*

The absorption spectra of low molecular weight gases are characterized by resolvable spectral features, as shown schematically in Fig. 3 (Kreuzer,

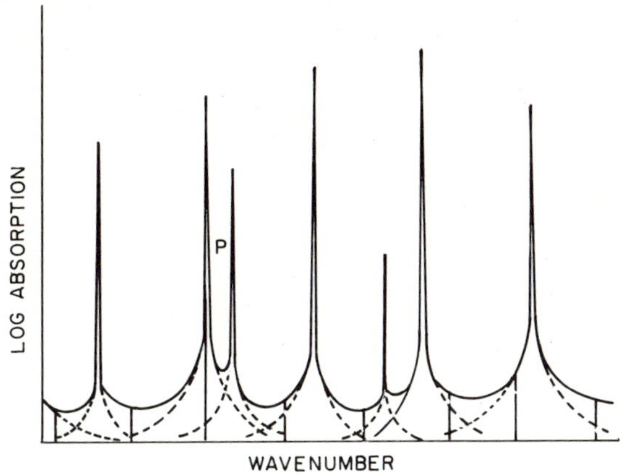

Fig. 3. Sampling of gas absorption at CO_2 laser emission wavelengths. (From Kreuzer, 1974, Fig. 2.)

1974). The solid curve, which represents the absorption coefficient of this hypothetical gas, is a superposition of several collision broadened lines, the tails of which are represented by the dashed lines. The narrow vertical lines represent the CO_2 laser emission wavelengths that occur in this spectral region and their height represents the absorption measured at each wavelength. Because of the discrete nature of both spectra one finds that there is one near coincidence, which is indicated by the point *P*.

Although the identification of this single coincidence between a laser wavelength and an absorption maximum may encourage the use of only that wavelength for detection and identification, caution must be exercised unless precise advance knowledge of the composition of the air sample is available. The reason for this becomes apparent on examination of Fig. 4, which is a plot of the absorption spectra of five vapors, as measured with a CO_2 laser (Kreuzer, 1974). Clearly, if the CO_2 laser wavelength that produces the largest optoacoustic signal from methanol (graph D) were chosen to detect and identify methanol, ethanol could be erroniously identified as methanol. On the other hand, if three or four additional laser wavelengths are also selected, the pattern of optoacoustic signals represents a "signature" of the methanol molecule, and identification errors can be reduced significantly.

An occasionally observed characteristic of the absorption spectra of light molecules is the presence of more than one resolvable absorption maximum within the tuning range of a single CO_2 laser transition. This type of feature has been observed in sulfur hexafluoride and in vinyl chloride, at concentrations in the parts per million range in nitrogen at 1 atm (Claspy, unpublished results). Such features increase the probability of accurate identification since the detailed structure is unlikely to be duplicated by another gas, nor can it be easily masked by an interfering species.

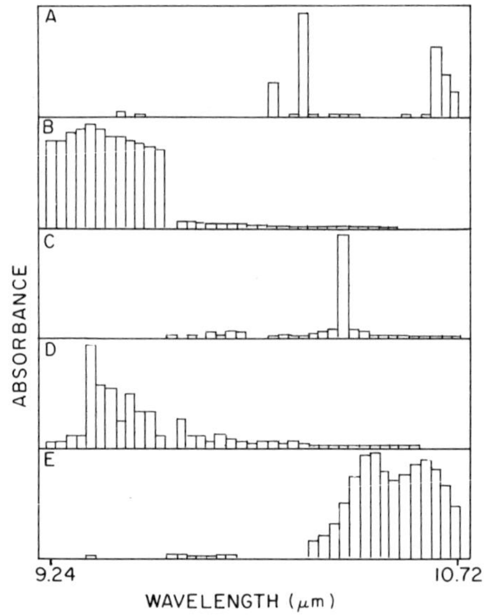

Fig. 4. Laser optoacoustic absorption spectra taken with prototype laser optoacoustic spectroscopy system at CO_2 laser wavelengths: (a) ammonia; (b) ethanol; (c) ethylene; (d) methanol; (e) trichloroethylene. (From Kreuzer, 1974, Fig. 4.)

E. Detection and Identification of Large-Molecule Gases

As pointed out in Section II.C, the approach to be used in the detection and identification of large-molecule vapors is somewhat different than that for small molecules. Here one must use the location of a broad spectral feature, frequently spanning several wavenumbers, for identification. For example, the —NO_2 group, which occurs in most explosives, exhibits an asymmetric stretch vibrational absorption band near 6 μm and an —O—N stretching mode near 11 μm, with the specific location being determined by the basis molecule to which the group is attached. Peak asymmetric stretch absorptions by —NO_2 occur at 6.039 and 6.086 μm in ethylene glycol dinitrate (EGDN) and at 5.970, 6.046, and 6.079 μm for nitroglycerine (NG). The peak —O—N absorptions occur at 11.148, 11.628, and 11.877 μm for EGDN and at 11.099 and 11.905 μm for NG (Pristera *et al.*, 1960).

While the use of spectral characteristics such as those described for EGDN and NG make detection identification of large molecules possible, the effects of interfering species can be much more serious than in the small-molecule problem.

F. Effect of Interfering Species

As indicated previously, the selection of measurement wavelengths for a specific vapor depends not only on the infrared absorption spectrum of that molecule, but also on the spectrum of any infrared-absorbing interfering species that may be present in the gas sample. Consider, as an example of this problem, the effect of water vapor on vapor detection and identification in the 5–7-μm spectral region.

The effect of water vapor interference has been considered theoretically by Gelbwachs (1974) and the effects have been observed experimentally by Claspy *et al.* (1976). Gelbwachs estimates that for the nearly 4000 water vapor absorption lines in the 5–7-μm region, fewer than 20% of the spectral intervals between lines exceed 9 GHz. In addition, he calculates that the limiting sensitivity for detection at 1 atm is $\sim 10^3$ and ~ 0.1 ppb in the proximity of strong and weak water vapor lines, respectively. Thus, great care must be exercised in the selection of identification wavelengths in the 5–7-μm spectral region.

Claspy and co-workers have observed that water vapor can completely mask the presence of small concentrations of such low vapor pressure large molecules as trinitrololuene and pentaerythrital tetranitrate. They suggest, furthermore, that water vapor entrained in or combined with the sample can mask the presence of a sample even when dry nitrogen (~ 1 ppm H_2O) is used as the carrier.

G. Saturation Effects

Since the absorption coefficient, and therefore the optoacoustic signal, is a function of incident optical power, consideration must be given to the effect of absorption saturation produced by excessive incident power. A description of this effect, based on a simple two-level model, is given below. In this model, the molecular system interacts resonantly with incident radiation and transitions between the two levels are stimulated. In addition, deexcitation from the upper level occurs because of fluorescence (spontaneous emission, with rate τ_R^{-1}) and also because of collisions (with rate τ_c^{-1}). The optoacoustic signal A is proportional to $N_2\tau_c^{-1}$, where N_2 is the number of molecules in the upper state, per unit volume.

Consider a two level system, and let N be the number of molecules $\text{cm}^{-3} = N_1 + N_2$, where N_1 is the number of molecules cm^{-3} in the lower state and N_2 the number of molecules cm^{-3} in the upper state. The rate equation description of the system may be written as

$$\frac{dN_1}{dt} = -B_{12}\frac{I_\nu h\nu}{c}N_1 + B_{21}\frac{I_\nu h\nu}{c}N_2 + (\tau_R^{-1} + \tau_c^{-1})N_2, \tag{5}$$

$$\frac{dN_2}{dt} = -B_{21}\frac{I_\nu h\nu}{c}N_2 + B_{12}\frac{I_\nu h\nu}{c}N_1 - (\tau_R^{-1} + \tau_c^{-1})N_2, \tag{6}$$

where $B_{12} = B_{21}$ and is Einstein's coefficient for stimulated emission and absorption, ν is the frequency of radiation interacting resonantly with the molecular system, I_ν the number of photons $\mathrm{cm}^{-1}\ \mathrm{sec}^{-1}$ of energy $h\nu$, τ_R the radiative lifetime of the upper state, and τ_c the collisional lifetime of the upper state.

For convenience these equations can be combined and slightly rewritten as

$$(d/dt)(N_2 - N_1) = -B(N_2 - N_1)I_\nu - 2\tau^{-1}N_2, \tag{7}$$

where

$$B = B_{12}h\nu/c, \qquad \tau^{-1} = (\tau_R^{-1} + \tau_c^{-1}), \qquad B_{12} = \frac{\pi^2 c^3}{\Delta\nu h\nu^3}A_{21} = \frac{\pi^2 c^3}{\Delta\nu h\nu^3 \tau_R}.$$

One may safely assume that steady-state conditions are always attained, that is, for any I_ν, the system has ample time to adjust itself to steady-state conditions appropriate to that value. This means that one can take

$$(d/dt)(N_2 - N_1) = 0,$$

and Eq. (3) yields

$$N_2 = BNI_\nu/(2BI_\nu + 2\tau^{-1}). \tag{8}$$

The acoustic signal is proportional to $N_2\tau_c^{-1}$, and accordingly the acoustic signal is given by

$$A \propto N_2\tau_c^{-1} = BNI_\nu\tau_c^{-1}/(2BI_\nu + 2\tau^{-1}). \tag{9}$$

In the case where the incident radiation is modulated at frequency ω so that

$$I_\nu = I_0(1 + \delta\sin\omega t), \qquad \text{where} \quad 0 \le \delta \le 1, \tag{10}$$

one can consistently write

$$N_2\tau_c^{-1} = \frac{BNI_0(1 + \delta\sin\omega t)\tau_c^{-1}}{2BI_0(1 + \delta\sin\omega t) + 2\tau^{-1}}. \tag{11}$$

Expanding in powers of ($\delta\sin\omega t$) and retaining only terms up to the second power of ($\delta \sin \omega t$),

$$N_2\tau_c^{-1} = \frac{BNI_0\tau_c^{-1}}{2BI_0 + 2\tau^{-1}} - \frac{4B^2I_0{}^2N\tau_c^{-1}\delta^2}{[2BI_0 + 2\tau^{-1}]^3} + \frac{2\tau_c^{-2}BNI_0}{(2BI_0 + 2\tau^{-1})^2}\delta \sin \omega t + \frac{4B^2I_0{}^2N\tau_c^{-1}}{(2BI_0 + 2\tau^{-1})^3}\delta^2 \sin 2\omega t + \cdots. \tag{12}$$

Insofar as the $\sin \omega t$ term is concerned, for very small I_0, $BI_0 \ll \tau^{-1}$ and

$$\frac{2\tau_c^{-2}BNI_0}{(2BI_0 + 2\tau^{-1})^2}\delta \to \frac{1}{2}\left(\frac{\tau}{\tau_c}\right)^2 BNI_0\delta. \tag{13}$$

But when I_0 becomes very large, then

$$\frac{2\tau_c^{-2}BNI_0}{(2BI_0 + 2\tau^{-1})^2}\delta \to \frac{1}{2}\tau_c^{-2}\frac{N}{BI_0}\delta. \tag{14}$$

It is quite transparent that BI_0 should be large but not larger than τ^{-1}. This can be demonstrated also by solving Eq. (10) for the optimum value of BI_0. Actually $BI_0 \equiv \tau^{-1}$ or the reciprocal of the lifetime as specified by interaction with radiation, and the optimum condition merely states (qualitatively) that one should not pump harder than the upper level can be collisionally depopulated.

Depending on the partial pressure of the absorbing gas, the presence or absence of other gases, and the influence of wall collisions, τ might be of the order of, say 10^{-6}–10^{-4} sec and it is found that saturation sets in at intensities of the order of fractions of a watt per square centimeter. This is in agreement with the results of Knox and Pao (1971) with $^{129}I_2$ and of Goldberg and Yusek (1970) with SF_6.

Furthermore the manner in which saturation occurs is of interest. In Eq. (12), the dominant dc term has I_0 to the same power in the numerator and in the denominator. Unnecessarily high values of I_0 might not bring correspondingly higher signal values, but there are no penalties either. This is different for the other terms. In particular, the ac term at frequency ω saturates according to I_0^{-1} and so does the second harmonic term. It is important therefore in practice to avoid using unnecessarily high incident powers.

Evidence of absorption saturation can be obtained by measurement of optoacoustic signal per unit optical power as a function of incident power, as shown schematically in Fig. 5. It is assumed here that the signal levels are sufficiently low that microphone response remains linear. If saturation effects are negligible, the normalized optoacoustic signal will not change as the incident power is increased. If, however, saturation effects are significant, the normalized signal decreases with increasing power.

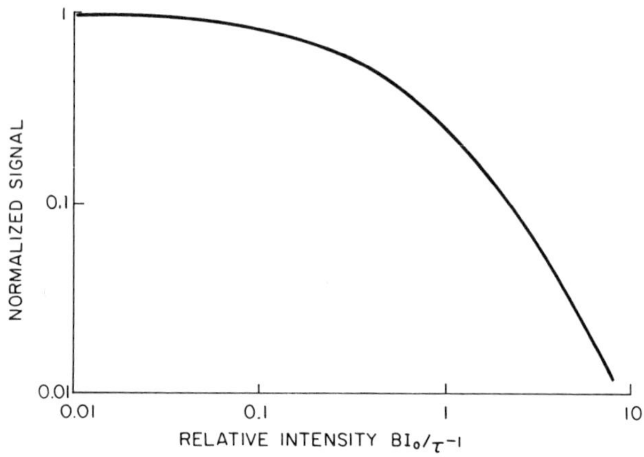

Fig. 5. Effect of incident power on normalized optoacoustic signal.

Pao and Claspy (1975) have reported the results of a study of saturation effects in nitroglycerine and found that saturation effects occurred at intensities of the order of 1 W cm^{-2}.

H. Analysis of Multicomponent Mixtures

One of the principle potential applications for optoacoustic detection system is that of detection, identification, and concentration measurement of hazardous industrial vapors in manufacturing plant environments. The atmosphere in such plants can contain a large number of vapors, many or all of which have specified maximum safe levels and must therefore be monitored. Since these safe levels range in general from about 100 ppb to about 100 ppm (Occupational Safety and Health Administration, 1972), accurate and reliable discrimination among them and measurement of individual concentrations is essential. A system that will perform this task must be calibrated for each gas, with identification wavelengths selected for minimal overlap among the constituents. The system must be capable of correlating spectral data obtained on site with calibration spectra so that accurate analysis can be done.

A multicomponent gas mixture can be analyzed by measuring the optoacoustic signal at each of a set of wavelengths chosen on the basis of the absorption spectra of the components. If the response of a particular optoacoustic cell at wavelength λ_i per unit concentration per unit incident power for a single component is R_i, then the signal S_i is

$$S_i = P_i C R_i, \tag{15}$$

where C is the concentration and P_i is the incident power. Then if the sample contains N components, the signal S_i is

$$S_i = P_i \sum_{n=1}^{N} R_{in} C_n, \tag{16}$$

where R_{in} is the response of component n at λ_i. Equation (16) shows that, by measuring the optoacoustic signal S_i and the incident optical power P_i at each wavelength selected for an N-component mixture, the concentration of each can be calculated by solving a set of simultaneous linear equations. The formal solution to Eq. (16) is

$$C_n = \sum_{i=1}^{N} R_{ni}^{-1}(S_i/P_i), \tag{17}$$

where R_{ni}^{-1} is the inverse of the matrix R_{in}.

The effectiveness of this method in the determination of concentrations of components in a gas mixture depends not only on the precision of measurement of S_i and P_i, but also on the nature of the matrix R_{in}. If a set of wavelengths can be selected so that R_{in} is diagonal, then the system is limited only by the noise level produced by the electronics and by scattering. Since such a choice of wavelengths is rarely, if ever, possible, consideration must be given to the ability of the system to discriminate among the various known components of the mixture as well as to spurious results arising from the presence of unexpected components. The effect of a large concentration of one gas A on the ability of a system to detect a small concentration of a second gas B can be described by a rejection ratio. The rejection ratio is defined as the concentration of gas A necessary to give the same signal as a unit concentration of gas B at the wavelength chosen for gas B. For example, Kreuzer (1974) reports that 1 ppm of ethanol can just be detected in the presence of 3200 ppm of ammonia at a wavelength of 9.25 μm. Other rejection ratios reported by Kreuzer are given in Table I (Kreuzer, 1974).

TABLE I

Rejection Ratios

Component being measured	Interfering component			
	Ethanol	Methanol	Ammonia	Trichloroethylene
Ethanol	—	270	3200	16,000
Methanol	760	—	1900	300
Ammonia	1080	430	—	1080
Trichloroethylene	200	200	1000	—

[a] From Kreuzer (1974, Table I).

The possibilities of incorrect identification or concentration measurement in gas mixtures, especially those resulting from the presence of unexpected gases, as may occur in industrial environments, can be minimized by the use of two or more wavelengths for the identification and measurement of each desired component. In this case concentrations can be calculated using Eq. (17) for each set of wavelengths and the results can be correlated to give a most probable identification and concentration.

III. Applications to Vapor Detection

The application of optoacoustic techniques to the detection of atmospheric pollutants has been reported by several authors (Kreuzer and Patel, 1971; Kreuzer *et al.*, 1972; Patel *et al.*, 1974). These include a study of NO concentration near a highway (Kreuzer and Patel, 1971), a study of the effect of solar ultraviolet radiation on the concentration of NO in the upper atmosphere (Patel *et al.*, 1974), and a determination of minimum detectable concentration of several common pollutants (Kreuzer *et al.*, 1972).

The study by Kreuzer and Patel (1971) of NO concentration near a highway demonstrates the need for care in the selection of the spectral feature to be used for identification and concentration measurement. The absorption spectrum of NO in the region from 1813 to 1823 cm^{-1} is shown in Fig. 6a, and consists of lines numbered 1, 5, 6, 8, and 11. Figure 6c is the spectrum of laboratory air at 21°C and 30% relative humidity, and indicates that there are strong water vapor absorptions on either extreme of the spectral region chosen for analysis. Clearly, then, for the NO measurement to be minimally affected, lines 6 and 8, which lie near the center of the interval, must be selected. Figures 6d and e are spectra taken at a roadside and near an automobile exhaust, respectively, and demonstrate measurement of ~2 ppm NO and 50 ppm NO in the presence of water vapor.

Kreuzer *et al.* (1972) have measured minimum sensitivities for detection of several atmospheric pollutants using a nonresonant optoacoustic cell. The results of these measurements are shown in Table II. CO and CO_2 lasers were used as excitation sources for these measurements, which were made at a sample pressure of 500 torr. The detection electronics included a lock-in amplifier with a 1-sec time constant and it was necessary to subtract a background signal, which resulted from absorption by and scattering from the cell windows.

In a paper by Patel *et al.* (1974), the results of a balloon-borne measurement of the concentration of NO as a function of altitude are reported. In

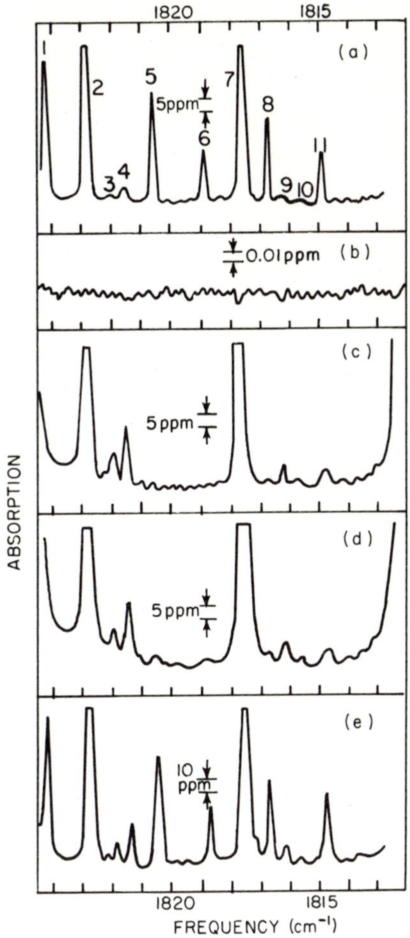

Fig. 6. Absorption spectra of various gas samples in the range of 1815–1825 cm^{-1} obtained by changing the magnetic field (of a spin-flip Raman laser) from 30 to 40 kg. All spectra are taken at a pressure of 500 torr. (a) Calibration spectrum of 20 ppm NO in N_2; lines numbered 1, 5, 6, 8, and 11 are due to NO; the remaining lines are due to H_2O; vertical sensitivity is 1.0; integration time is 1 sec. (b) Absorption cell noise; vertical sensi tivity is 100; integration time is 4 secs. (c) Room air at 21°C and 30% relative humidity; vertical sensitivity is 1; integration time is 1 sec. (d) Air sample from Route 22 near Plainfield, N.J.; vertical sensitivity is 1; integration time is 1 sec. (E) Automobile exhaust; vertical sensitivity is 0.4; integration time is 1 sec. (From Kreuzer and Patel, 1971, Fig. 2. Copyright 1971 by the American Association for the Advancement of Science.)

this experiment two NO absorptions near 1887 cm^{-1} were selected for identification and a water vapor line near 1889 cm^{-1} was selected for H_2O concentration measurement. Figures 7a and b are spectra of air samples taken at 28-km altitude before and after ultraviolet sunrise, respectively. The NO identification was on the basis of the widths of the two lines in Fig. 7b, with that at 1887.55 cm^{-1} being due to two Λ-doubling split $\Omega = \frac{1}{2}$ lines that have been pressure broadened into a single feature of about twice the width of the single line at 1887.63 cm^{-1}. Figure 7a shows no structure and therefore indicates a maximum NO concentration of $\sim 1.5 \times 10^8$ cm^{-3} in the absence of sunlight. Figure 7b shows the result of a spectrum taken $\sim$5.5 hr after sunrise, at which time the spectrum showed two distinct

TABLE II

Noise-Limited Sensitivities for Detecting Pollutant Gases[a]

Gas	Sensitivity (ppb)	Laser	Infrared source transition	Wavelength (μm)
Ammonia	0.4	CO	$P_{19-18}(15)$	6.1493
Benzene	3.0	CO_2	00°1–02°0 P(30)	9.6392
1,3-Butadiene	1.0	CO	$P_{20-19}(13)$	6.2153
1,3-Butadiene	2.0	CO_2	00°1–10°0 P(30)	10.6964
1-Butene	2.0	CO	$P_{19-18}(9)$	6.0685
1-Butene	2.0	CO_2	00°1–10°0 P(38)	10.7874
Ethylene	0.2	CO_2	00°1–10°0 P(14)	10.5321
Methanol	0.3	CO_2	00°1–02°0 P(34)	9.6760
Nitric oxide	0.4	CO	$P_{8-7}(11)$	5.2148
Nitrogen dioxide	0.1	CO	$P_{20-19}(14)$	6.2293
Propylene	3.0	CO	$P_{19-18}(9)$	6.0685
Trichloroethylene	0.7	CO_2	00°1–10°0 P(24)	10.6321
Water	14.0	CO	$P_{17-16}(13)$	5.9417

[a] From Kreuzer *et al.* (1972, Table I). Copyright 1972 by the American Association for the Advancement of Science.

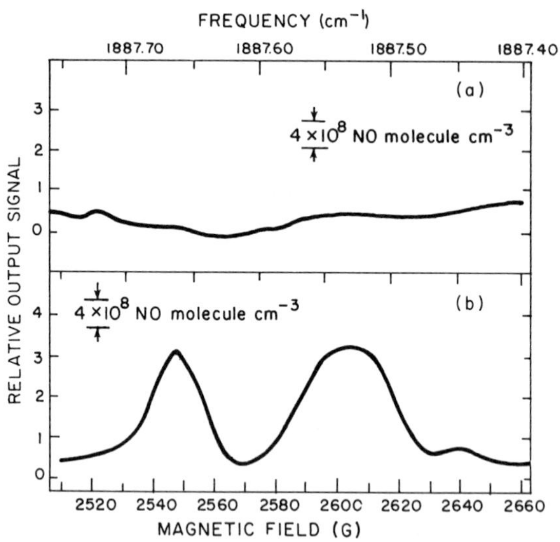

Fig. 7. (a) Relative output signal as a function of the spin-flip Raman (SFR) laser magnetic field for NO, taken before ultra-violet sunrise (11:55:00 to 12:27:00 GMT, 19 October 1973). The NO absorption signals are expected to occur at magnetic fields of 2545 and 2605 G. Calibration in equivalent NO molecules per cubic centimeter is given. (b) Relative output signal as a function of the SFR laser magnetic field for NO taken after ultraviolet sunrise (17:21:50 to 17:52:05 GMT, 19 October 1973); NO absorption signals at magnetic fields of 2545 and 2605 G can be seen. Calibration in equivalent NO molecules per cubic centimeter is given, indicating a NO concentration of $(2 \pm 0.15) \times 10^9$ molecule cm^{-3}. (Data taken at 28-km altitude and 225°K temperature.) (From Patel *et al.*, 1974, Fig. 1. Copyright 1974 by the American Association for the Advancement of Science.)

TABLE III

Minimum Detectable Explosive Concentration Using Four Wavelengths[a]

Wavelength (μm)	Explosive	Minimum Detectable Concentration (ppm vapor)
6	NG	0.24×10^{-1}
	EGDN	2.5
	DNT	220.0
9	NG	0.055×10^{-2}
	EGDN	2.5
	DNT	16.0
11	NG	0.028×10^{-2}
	EGDN	1.5

[a] From Claspy *et al.* (1976, Table IV).

peaks corresponding to NO absorptions. The measured NO concentration at that time was $2 \pm 0.15 \times 10^{19}\,\text{cm}^{-3}$.

A summary of the results of a feasibility study for the detection of nitroglycerine, dinitrotoluene, and ethylene glycol dinitrate, by Claspy *et al.* (1976) is shown in Table III. In this experiment, which involves the detection of large molecules, a technique like that discussed in Section II was used. In each spectral region, namely, near 6, 9.4, and 11 μm, the optoacoustic signal was determined at each of several different laser transitions. It was assumed that measurement on the four most strongly absorbed lines is sufficient for detection and identification of the vapor. Although the minimum detectable concentration per milliwatt of laser power is nearly the same in each spectral region when dry N_2 is used as the carrier gas, the situation becomes significantly more difficult at 6 μm when air with a 50% relative humidity is used. In this case the water vapor signal masks the explosive signal so effectively that detection is virtually impossible at 6 μm. Detection at 9.4 and at 11 μm is unaffected by the water vapor.

IV. Application of Optoacoustic Techniques to the Measurement of Molecular Relaxation Rates

The measurement of molecular relaxation rates by use of optoacoustic techniques has been discussed by numerous authors. In a theoretical discussion which assumes that the incident intensity is much less than the saturation intensity, Kaiser (1959) shows that the optoacoustic signal is given by

$$S_{\text{fund}} = -\frac{K\tau^2}{\omega\tau(1+\omega^2\tau^2)^{1/2}}\cos(\omega t - \phi) \tag{18}$$

where S_{fund} is the optoacoustic signal at the chopping frequency ω, τ is the excited state relaxation time, K is a proportionality constant, and the phase angle ϕ between the chopper and the optoacoustic signal is

$$\tan\phi = \omega\tau. \tag{19}$$

The dependence of the amplitude and phase of S_{fund} on the quantity $\omega\tau$ is shown in Figs. 8 and 9, respectively.

This technique has received some attention since it was first suggested by Slobodskaya (1948). Cottrell *et al.* (1966) made use of both the signal amplitude effect and the phase change effect to study relaxation rates in CO_2, CH_4, N_2O, NH_3HCN, and $CClF_3$ and obtained results that were in general agreement with those obtained by ultrasonic methods.

Cottrell *et al.* (1966), however, point out two significant factors which must be considered in optoacoustic measurement of molecular relaxations. First, since frequency response of S_{fund} or ϕ is the quantity being measured, the optoacoustic cell must be nonresonant over the frequency range to be

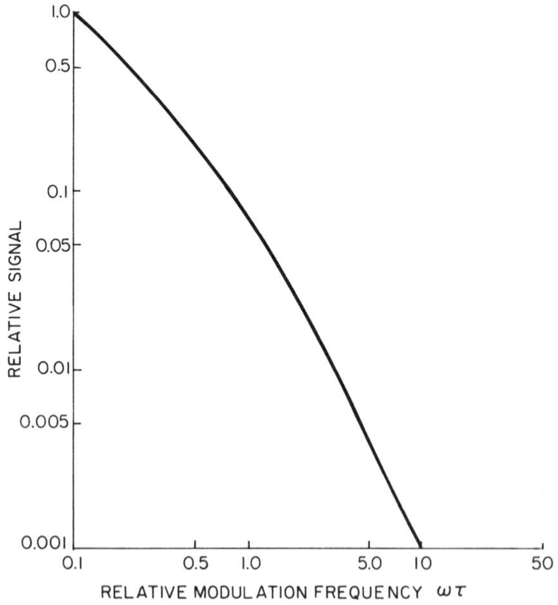

Fig. 8. Relative amplitude of the fundamental frequency signal component as a function of the relative modulation frequency $\omega\tau$.

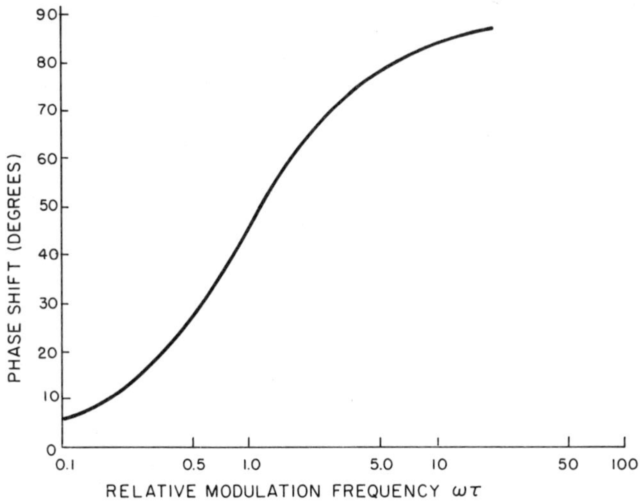

Fig. 9. Phase angle of the fundamental frequency signal component as a function of the relative modulation frequency $\omega\tau$.

studied. Second, what is observed in the optoacoustic detector is not the decay of vibrational energy from a specific mode, but the appearance of translational energy in a gas that has been excited to a specific vibrational mode by the absorption of radiation. Therefore, the relaxation time observed by this method may not be the average lifetime of this mode, but just a measure of the time for which the energy originally put into this mode remains locked up in the molecule, perhaps in a variety of coupled modes. Therefore, all possible internal resonances must be investigated before conclusions may be drawn.

V. Conclusion

Optoacoustic spectroscopic techniques provide a sensitive, accurate method for identification and detection of vapors in the atmosphere, provided that care is taken to select features of the absorption spectrum of the molecule being sought that are not easily masked by overlapping spectra of other molecular species which may be present in the system. Sensitivities in the parts per billion partial pressure ratio are easily attained for many atmospheric pollutants and industrial vapors.

Potential future applications include monitoring of concentrations of each of several industrial vapors in a manufacturing environment. Here the cell could be separately calibrated for each vapor and the signatures stored in a small computer. Detection, identification, and measurement would then be accomplished by comparison of measured signals with those stored in the memory while the laser is automatically stepped through the appropriate spectral region.

Appendix: CO_2 and CO Laser Wavelengths

The most readily available discretely tunable infrared laser sources for optoacoustic spectroscopy are the CO_2 and CO lasers. Accurate measurements of wavelengths available from these sources have been reported in the literature and are given here for convenience of the reader.

A. CO_2 Laser Wavelengths

Chang (1970) has calculated wave numbers and wavelengths of 120 laser lines of the $^{12}C^{16}O_2$ laser using the results of absolute frequency measurements and the results of accurate measurements of rotational constants. These results are given in Tables A1 and A2. The vacuum wave numbers were calculated by Chang using a value of 2.997925×10^{10} cm sec^{-1} for the velocity of light and the dry air wavelengths were calculated by him using 1.0002726 for the refractive index of air.

Freed *et al.* (1974) have reported vacuum wave numbers and wavelengths for laser lines of several isotopic variations of the CO_2 laser. These results were calculated from the results of accurate difference frequency measurements between the isotope laser lines and lines from a stabilized $^{12}C^{16}O_2$ reference laser. The results for $^{12}C^{18}O_2$ are given in Tables A3 and A4, those for $^{13}C^{16}O_2$ in Tables A5 and A6, and those for $^{13}C^{18}O_2$ in Tables A7 and A8. Dry air wavelengths were calculated by this author using the wave numbers reported by Freed *et al.* (1974), and the refractive index calculations of Coleman *et al.* (1960).

B. CO Laser Wavelengths

The emission spectrum of the CO laser has been measured relative to the visible atomic lines of neon by Mantz *et al.* (1970), and by Roh and Rao (1974). The results of these observations are given in Table A9. Dry air wavelengths were calculated by the present author using the wave numbers reported by Mantz *et al* (1970) and by Roh and Rao (1974), and the refractive index calculations of Coleman *et al.* (1960).

TABLE A1

Wavenumbers and Wavelengths for the $^{12}C^{16}O_2$ Laser $00^{\circ}1$–$[10^{\circ}0, 02^{\circ}0]_{I}$ Band[a]

LINE	WV.N. (INV. CM.)	MICRONS(VAC)	MICRONS(AIR)
P(60)	903.2174	11.07153	11.06851
P(58)	905.5111	11.04348	11.04047
P(56)	907.7782	11.01590	11.01290
P(54)	910.0188	10.98878	10.98578
P(52)	912.2331	10.96211	10.95912
P(50)	914.4212	10.93587	10.93289
P(48)	916.5833	10.91008	10.90710
P(46)	918.7196	10.88471	10.88174
P(44)	920.8302	10.85976	10.85680
P(42)	922.9152	10.83523	10.83227
P(40)	924.9748	10.81110	10.80815
P(38)	927.0091	10.78738	10.78444
P(36)	929.0182	10.76405	10.76111
P(34)	931.0021	10.74111	10.73818
P(32)	932.9611	10.71856	10.71563
P(30)	934.8952	10.69638	10.69347
P(28)	936.8044	10.67458	10.67167
P(26)	938.6889	10.65315	10.65025
P(24)	940.5488	10.63209	10.62919
P(22)	942.3840	10.61138	10.60849
P(20)	944.1947	10.59103	10.58814
P(18)	945.9809	10.57103	10.56815
P(16)	947.7427	10.55138	10.54851
P(14)	949.4800	10.53208	10.52921
P(12)	951.1929	10.51311	10.51024
P(10)	952.8815	10.49448	10.49162
P(8)	954.5458	10.47618	10.47333
P(6)	956.1856	10.45822	10.45536
P(4)	957.8012	10.44058	10.43773
P(2)	959.3924	10.42326	10.42042
P(0)	960.9592	10.40626	10.40343
V(0)	961.7335	10.39789	10.39505
R(2)	963.2638	10.38137	10.37854
R(4)	964.7696	10.36516	10.36234
R(6)	966.2510	10.34927	10.34645
R(8)	967.7079	10.33369	10.33088
R(10)	969.1402	10.31842	10.31561
R(12)	970.5479	10.30345	10.30065
R(14)	971.9309	10.28879	10.28599
R(16)	973.2892	10.27443	10.27163
R(18)	974.6226	10.26038	10.25758
R(20)	975.9311	10.24662	10.24383
R(22)	977.2146	10.23316	10.23037
R(24)	978.4729	10.22000	10.21722
R(26)	979.7061	10.20714	10.20436
R(28)	980.9139	10.19457	10.19179
R(30)	982.0962	10.18230	10.17952
R(32)	983.2529	10.17032	10.16755
R(34)	984.3839	10.15863	10.15586
R(36)	985.4891	10.14724	10.14448
R(38)	986.5682	10.13614	10.13338
R(40)	987.6215	10.12533	10.12257
R(42)	988.6477	10.11482	10.11207
R(44)	989.6477	10.10460	10.10185
R(46)	990.6211	10.09467	10.09192
R(48)	991.5676	10.08504	10.08229
R(50)	992.4870	10.07569	10.07295
R(52)	993.3791	10.06665	10.06390
R(54)	994.2437	10.05789	10.05515
R(56)	995.0807	10.04943	10.04669
R(58)	995.8897	10.04127	10.03853

[a] From Chang (1970, Table 1). Frequencies in megahertz given by Chang are not included here.

TABLE A2

Wavenumbers and Wavelengths for the $^{12}C^{16}O_2$ Laser $00°1-[10°0, 02°0]_{II}$ Band[a]

LINE	WV.N. (INV. CM.)	MICRONS(VAC)	MICRONS(AIR)
P(60)	1005.4840	9.945459	9.942748
P(58)	1007.7765	9.922835	9.920131
P(56)	1010.0468	9.900531	9.897832
P(54)	1012.2948	9.878544	9.875852
P(52)	1014.5202	9.856876	9.854189
P(50)	1016.7227	9.835523	9.832843
P(48)	1018.9019	9.814487	9.811812
P(46)	1021.0578	9.793765	9.791095
P(44)	1023.1900	9.773356	9.770692
P(42)	1025.2982	9.753259	9.750601
P(40)	1027.3824	9.733474	9.730821
P(38)	1029.4422	9.713998	9.711350
P(36)	1031.4775	9.694831	9.692188
P(34)	1033.4880	9.675971	9.673333
P(32)	1035.4736	9.657416	9.654784
P(30)	1037.4314	9.639191	9.636564
P(28)	1039.3693	9.621219	9.618597
P(26)	1041.2790	9.603573	9.600956
P(24)	1043.1632	9.586227	9.583615
P(22)	1045.0216	9.569179	9.566571
P(20)	1046.8542	9.552428	9.549825
P(18)	1048.6608	9.535972	9.533373
P(16)	1050.4413	9.519808	9.517214
P(14)	1052.1955	9.503937	9.501346
P(12)	1053.9235	9.488355	9.485769
P(10)	1055.6250	9.473060	9.470478
P(8)	1057.3001	9.458052	9.455474
P(6)	1058.9487	9.443328	9.440754
P(4)	1060.5706	9.428886	9.426316
P(2)	1062.1659	9.414725	9.412159
P(0)	1063.7345	9.400842	9.398280
V(0)	1064.5086	9.394004	9.391444
R(2)	1066.0373	9.380534	9.377978
R(4)	1067.5390	9.367339	9.364786
R(6)	1069.0140	9.354414	9.351864
R(8)	1070.4622	9.341758	9.339212
R(10)	1071.8837	9.329370	9.326827
R(12)	1073.2784	9.317247	9.314707
R(14)	1074.6464	9.305386	9.302850
R(16)	1075.9877	9.293786	9.291253
R(18)	1077.3024	9.282444	9.279914
R(20)	1078.5906	9.271358	9.268831
R(22)	1079.8522	9.260526	9.258002
R(24)	1081.0874	9.249946	9.247425
R(26)	1082.2962	9.239615	9.237096
R(28)	1083.4787	9.229530	9.227015
R(30)	1084.6351	9.219690	9.217177
R(32)	1085.7654	9.210092	9.207582
R(34)	1086.8698	9.200733	9.198226
R(36)	1087.9484	9.191612	9.189107
R(38)	1089.0013	9.182725	9.180222
R(40)	1090.0287	9.174070	9.171570
R(42)	1091.0307	9.165644	9.163146
R(44)	1092.0075	9.157446	9.154950
R(46)	1092.9593	9.149471	9.146978
R(48)	1093.8862	9.141719	9.139227
R(50)	1094.7885	9.134184	9.131695
R(52)	1095.6662	9.126867	9.124379
R(54)	1096.5198	9.119762	9.117277
R(56)	1097.3493	9.112868	9.110384
R(58)	1098.1550	9.106182	9.103700

[a] From Chang (1970, Table 2). Frequencies in megahertz by Chang are not included here.

TABLE A3

Calculated Wavenumbers and Wavelengths of the $^{12}C^{18}O_2$ Laser 00°1–[10°0, 02°0]$_I$ Band[a]

LINE	WV.N. (INV. CM.)	MICRONS(VAC)	MICRONS(AIR)
P(60)	913.4618	10.94736	10.94438
P(58)	915.6390	10.92133	10.91635
P(56)	917.7882	10.89575	10.89279
P(54)	919.9095	10.87063	10.86767
P(52)	922.0032	10.84594	10.84299
P(50)	924.0691	10.82170	10.81875
P(48)	926.1076	10.79788	10.79493
P(46)	928.1186	10.77448	10.77154
P(44)	930.1022	10.75150	10.74857
P(42)	932.0586	10.72893	10.72601
P(40)	933.9879	10.70677	10.70385
P(38)	935.8900	10.68501	10.68210
P(36)	937.7651	10.66365	10.66074
P(34)	939.6133	10.64267	10.63977
P(32)	941.4347	10.62208	10.61919
P(30)	943.2292	10.60187	10.59898
P(28)	944.9969	10.58204	10.57916
P(26)	946.7380	10.56258	10.55970
P(24)	948.4524	10.54349	10.54061
P(22)	950.1401	10.52476	10.52189
P(20)	951.8013	10.50639	10.50353
P(18)	953.4359	10.48838	10.48552
P(16)	955.0440	10.47072	10.46786
P(14)	956.6256	10.45340	10.45056
P(12)	958.1807	10.43644	10.43359
P(10)	959.7093	10.41982	10.41698
P(8)	961.2115	10.40353	10.40070
P(6)	962.6872	10.38758	10.38475
P(4)	964.1364	10.37197	10.36914
P(2)	965.5591	10.35669	10.35387
P(0)	966.9554	10.34173	10.33891
V(0)	967.6436	10.33438	10.33156
R(2)	969.0001	10.31991	10.31710
R(4)	970.3300	10.30577	10.30296
R(6)	971.6334	10.29194	10.28914
R(8)	972.9101	10.27844	10.27563
R(10)	974.1603	10.26525	10.26245
R(12)	975.3837	10.25237	10.24958
R(14)	976.5804	10.23981	10.23702
R(16)	977.7503	10.22755	10.22477
R(18)	978.8933	10.21561	10.21283
R(20)	980.0095	10.20398	10.20120
R(22)	981.0986	10.19265	10.18987
R(24)	982.1607	10.18163	10.17885
R(26)	983.1957	10.17091	10.16814
R(28)	984.2034	10.16050	10.15773
R(30)	985.1836	10.15038	10.14762
R(32)	986.1369	10.14057	10.13781
R(34)	987.0624	10.13107	10.12830
R(36)	987.9604	10.12186	10.11910
R(38)	988.8306	10.11295	10.11019
R(40)	989.6730	10.10434	10.10159
R(42)	990.4875	10.09603	10.09328
R(44)	991.2740	10.08802	10.08527
R(46)	992.0322	10.08031	10.07756
R(48)	992.7621	10.07290	10.07016
R(50)	993.4636	10.06579	10.06304
R(52)	994.1365	10.05898	10.05623
R(54)	994.7807	10.05246	10.04972
R(56)	995.3959	10.04625	10.04351
R(58)	995.9821	10.04034	10.03760

[a] From Freed *et al.* (1974, Table VI). Frequencies in megahertz given by Freed *et al.* are not included here. Dry air wavelengths were calculated from the above data by this author.

TABLE A4

Calculated Wavenumbers and Wavelengths of the $^{12}C^{18}O_2$ Laser $00^\circ1-[10^\circ0, 02^\circ0]_{II}$ Band[a]

LINE	WV.N. (INV. CM.)	MICRONS(VAC)	MICRONS(AIR)
P(60)	1033.7729	9.6733040	9.6706676
P(58)	1035.7000	9.6553055	9.6526740
P(56)	1037.6100	9.6375315	9.6349049
P(54)	1039.5030	9.6199812	9.6173594
P(52)	1041.3787	9.6026539	9.6000368
P(50)	1043.2370	9.5855488	9.5829363
P(48)	1045.0778	9.5686651	9.5660572
P(46)	1046.9009	9.5520019	9.5493985
P(44)	1048.7062	9.5355584	9.5329596
P(42)	1050.4936	9.5193338	9.5167393
P(40)	1052.2630	9.5033271	9.5007370
P(38)	1054.0142	9.4875374	9.4849517
P(36)	1055.7472	9.4719639	9.4693824
P(34)	1057.4618	9.4566056	9.4540283
P(32)	1059.1580	9.4414615	9.4388883
P(30)	1060.8356	9.4265307	9.4239616
P(28)	1062.4946	9.4118121	9.4092470
P(26)	1064.1348	9.3973048	9.3947437
P(24)	1065.7563	9.3830078	9.3804505
P(22)	1067.3588	9.3689199	9.3663665
P(20)	1068.9424	9.3550402	9.3524906
P(18)	1070.5070	9.3413676	9.3388217
P(16)	1072.0525	9.3279011	9.3253588
P(14)	1073.5788	9.3146394	9.3121008
P(12)	1075.0859	9.3015815	9.2990465
P(10)	1076.5738	9.2887264	9.2861948
P(8)	1078.0424	9.2760729	9.2735447
P(6)	1079.4916	9.2636197	9.2610950
P(4)	1080.9214	9.2513659	9.2488445
P(2)	1082.3318	9.2393102	9.2367921
P(0)	1083.7228	9.2274514	9.2249365
V(0)	1084.4110	9.2215955	9.2190822
R(2)	1085.7728	9.2100299	9.2075197
R(4)	1087.1150	9.1986582	9.1961512
R(6)	1088.4376	9.1874793	9.1849753
R(8)	1089.7410	9.1764918	9.1739908
R(10)	1091.0247	9.1656946	9.1631966
R(12)	1092.2889	9.1550865	9.1525913
R(14)	1093.5336	9.1446661	9.1421738
R(16)	1094.7587	9.1344321	9.1319426
R(18)	1095.9644	9.1243834	9.1218966
R(20)	1097.1506	9.1145185	9.1120344
R(22)	1098.3173	9.1048363	9.1023548
R(24)	1099.4646	9.0953353	9.0928565
R(26)	1100.5925	9.0860143	9.0835380
R(28)	1101.7011	9.0768719	9.0743981
R(30)	1102.7903	9.0679068	9.0654354
R(32)	1103.8602	9.0591176	9.0566486
R(34)	1104.9109	9.0505029	9.0480363
R(36)	1105.9424	9.0420614	9.0395971
R(38)	1106.9549	9.0337917	9.0313296
R(40)	1107.9482	9.0256923	9.0232324
R(42)	1108.9226	9.0177619	9.0153041
R(44)	1109.8780	9.0099989	9.0075433
R(46)	1110.8146	9.0024021	8.9999486
R(48)	1111.7324	8.9949699	8.9925184
R(50)	1112.6315	8.9877009	8.9852514
R(52)	1113.5121	8.9805936	8.9781460
R(54)	1114.3741	8.9736464	8.9712007
R(56)	1115.2178	8.9668580	8.9644142
R(58)	1116.0431	8.9602268	8.9577848

[a] From Freed *et al.* (1974, Table VII). Frequencies in megahertz given by Freed *et al.* are not included here. Dry air wavelengths were calculated from the above data by this author.

TABLE A5

Calculated Wavenumbers and Wavelengths of the $^{13}C^{16}O_2$ Laser $00^01-[10^00, 02^00]_I$ Band[a]

LINE	WV.N. (INV. CM.)	MICRONS(VAC)	MICRONS(AIR)
P(60)	857.9473	11.65572	11.65255
P(58)	860.0889	11.62670	11.62353
P(56)	862.2096	11.59810	11.59494
P(54)	864.3095	11.56992	11.56677
P(52)	866.3886	11.54216	11.53901
P(50)	868.4471	11.51480	11.51166
P(48)	870.4851	11.48784	11.48471
P(46)	872.5027	11.46128	11.45815
P(44)	874.5000	11.43510	11.43198
P(42)	876.4770	11.40931	11.40620
P(40)	878.4339	11.38389	11.38079
P(38)	880.3707	11.35885	11.35575
P(36)	882.2875	11.33417	11.33108
P(34)	884.1844	11.30985	11.30677
P(32)	886.0614	11.28589	11.28282
P(30)	887.9186	11.26229	11.25922
P(28)	889.7561	11.23903	11.23597
P(26)	891.5738	11.21612	11.21306
P(24)	893.3719	11.19354	11.19049
P(22)	895.1504	11.17130	11.16826
P(20)	896.9093	11.14939	11.14636
P(18)	898.6486	11.12781	11.12478
P(16)	900.3685	11.10656	11.10353
P(14)	902.0687	11.08562	11.08260
P(12)	903.7495	11.06501	11.06199
P(10)	905.4108	11.04470	11.04169
P(8)	907.0526	11.02471	11.02171
P(6)	908.6749	11.00503	11.00203
P(4)	910.2777	10.98565	10.98266
P(2)	911.8610	10.96658	10.96359
P(0)	913.4248	10.94780	10.94482
V(0)	914.1994	10.93853	10.93555
R(2)	915.7338	10.92020	10.91722
R(4)	917.2486	10.90216	10.89919
R(6)	918.7439	10.88442	10.88145
R(8)	920.2194	10.86697	10.86401
R(10)	921.6752	10.84980	10.84685
R(12)	923.1113	10.83292	10.82997
R(14)	924.5276	10.81633	10.81338
R(16)	925.9239	10.80002	10.79707
R(18)	927.3004	10.78399	10.78105
R(20)	928.6568	10.76824	10.76530
R(22)	929.9931	10.75276	10.74983
R(24)	931.3093	10.73757	10.73464
R(26)	932.6052	10.72265	10.71972
R(28)	933.8807	10.70800	10.70508
R(30)	935.1359	10.69363	10.69071
R(32)	936.3705	10.67953	10.67662
R(34)	937.5844	10.66570	10.66279
R(36)	938.7776	10.65214	10.64924
R(38)	939.9499	10.63886	10.63596
R(40)	941.1013	10.62584	10.62295
R(42)	942.2315	10.61310	10.61021
R(44)	943.3404	10.60062	10.59773
R(46)	944.4280	10.58841	10.58553
R(48)	945.4941	10.57648	10.57359
R(50)	946.5385	10.56481	10.56193
R(52)	947.5610	10.55340	10.55053
R(54)	948.5616	10.54227	10.53940
R(56)	949.5401	10.53141	10.52854
R(58)	950.4962	10.52081	10.51795

[a] From Freed *et al.* (1974, Table VIII). Frequencies in megahertz given by Freed *et al.* are not included here. Dry air wavelengths were calculated from the above data by this author.

TABLE A6

Calculated Wavenumbers and Wavelengths of the $^{13}C^{16}O_2$ Laser 00°1–[10°0, 02°0]$_{II}$ Band[a]

LINE	WV.N. (INV. CM.)	MICRONS(VAC)	MICRONS(AIR)
P(60)	958.3085	10.43505	10.43220
P(58)	960.6708	10.40939	10.40655
P(56)	963.0087	10.38412	10.38129
P(54)	965.3222	10.35923	10.35641
P(52)	967.6108	10.33473	10.33191
P(50)	969.8743	10.31061	10.30780
P(48)	972.1123	10.28687	10.28407
P(46)	974.3247	10.26351	10.26072
P(44)	976.5112	10.24053	10.23774
P(42)	978.6715	10.21793	10.21514
P(40)	980.8055	10.19570	10.19292
P(38)	982.9127	10.17384	10.17106
P(36)	984.9931	10.15235	10.14958
P(34)	987.0465	10.13123	10.12847
P(32)	989.0726	10.11048	10.10772
P(30)	991.0713	10.09009	10.08734
P(28)	993.0423	10.07006	10.06731
P(26)	994.9855	10.05039	10.04765
P(24)	996.9008	10.03108	10.02835
P(22)	998.7880	10.01213	10.00940
P(20)	1000.6470	9.99353	9.99081
P(18)	1002.4775	9.97528	9.97256
P(16)	1004.2796	9.95738	9.95467
P(14)	1006.0531	9.93983	9.93712
P(12)	1007.7979	9.92262	9.91991
P(10)	1009.5138	9.90575	9.90305
P(8)	1011.2009	9.88923	9.88653
P(6)	1012.8591	9.87304	9.87035
P(4)	1014.4882	9.85718	9.85450
P(2)	1016.0883	9.84166	9.83898
P(0)	1017.6592	9.82647	9.82379
V(0)	1018.4338	9.81899	9.81632
R(2)	1019.9610	9.80429	9.80162
R(4)	1021.4591	9.78991	9.78724
R(6)	1022.9280	9.77585	9.77319
R(8)	1024.3677	9.76211	9.75945
R(10)	1025.7782	9.74869	9.74603
R(12)	1027.1596	9.73558	9.73293
R(14)	1028.5119	9.72278	9.72013
R(16)	1029.8351	9.71029	9.70764
R(18)	1031.1292	9.69810	9.69546
R(20)	1032.3944	9.68621	9.68357
R(22)	1033.6307	9.67463	9.67199
R(24)	1034.8382	9.6633460	9.6607124
R(26)	1036.0169	9.6523516	9.6497209
R(28)	1037.1670	9.6416483	9.6390206
R(30)	1038.2886	9.6312334	9.6286085
R(32)	1039.3817	9.6211040	9.6184818
R(34)	1040.4466	9.6112571	9.6086377
R(36)	1041.4833	9.6016899	9.5990731
R(38)	1042.4920	9.5923994	9.5897851
R(40)	1043.4726	9.5833826	9.5807707
R(42)	1044.4260	9.5746365	9.5720270
R(44)	1045.3517	9.5661580	9.5635508
R(46)	1046.2500	9.5579441	9.5553392
R(48)	1047.1213	9.5499917	9.5473889
R(50)	1047.9656	9.5422977	9.5396970
R(52)	1048.7832	9.5348588	9.5322601
R(54)	1049.5743	9.5276719	9.5250752
R(56)	1050.3392	9.5207337	9.5181389
R(58)	1051.0780	9.5140410	9.5114480

[a] From Freed *et al.* (1974, Table IX). Frequencies in megahertz given by Freed *et al.* are not included here. Dry air wavelengths were calculated from the above data by this author.

TABLE A7

Calculated Wavenumbers and Wavelengths of the $^{13}C^{18}O_2$ Laser 00°1–[10°0, 02°0]$_I$ Band[a]

LINE	WV.N. (INV. CM.)	MICRONS(VAC)	MICRONS(AIR)
P(60)	876.8831	11.40402	11.40091
P(58)	878.9457	11.37726	11.37416
P(56)	880.9839	11.35094	11.34785
P(54)	882.9979	11.32505	11.32196
P(52)	884.9879	11.29958	11.29650
P(50)	886.9539	11.27454	11.27146
P(48)	888.8962	11.24990	11.24684
P(46)	890.8147	11.22567	11.22261
P(44)	892.7097	11.20184	11.19879
P(42)	894.5812	11.17841	11.17536
P(40)	896.4294	11.15536	11.15232
P(38)	898.2543	11.13270	11.12967
P(36)	900.0560	11.11041	11.10739
P(34)	901.8346	11.08850	11.08548
P(32)	903.5902	11.06696	11.06394
P(30)	905.3229	11.04578	11.04277
P(28)	907.0327	11.02495	11.02195
P(26)	908.7197	11.00449	11.00149
P(24)	910.3840	10.98437	10.98138
P(22)	912.0256	10.96460	10.96161
P(20)	913.6445	10.94517	10.94219
P(18)	915.2408	10.92608	10.92310
P(16)	916.8145	10.90733	10.90435
P(14)	918.3657	10.88890	10.88593
P(12)	919.8944	10.87081	10.86784
P(10)	921.4005	10.85304	10.85008
P(8)	922.8847	10.83558	10.83263
P(6)	924.3454	10.81846	10.81551
P(4)	925.7841	10.80165	10.79871
P(2)	927.2003	10.78515	10.78221
P(0)	928.5940	10.76896	10.76603
V(0)	929.2824	10.76099	10.75805
R(2)	930.6424	10.74526	10.74233
R(4)	931.9797	10.72984	10.72692
R(6)	933.2945	10.71473	10.71181
R(8)	934.5867	10.69991	10.69699
R(10)	935.8563	10.68540	10.68248
R(12)	937.1031	10.67118	10.66827
R(14)	938.3272	10.65726	10.65435
R(16)	939.5284	10.64363	10.64073
R(18)	940.7068	10.63030	10.62740
R(20)	941.8623	10.61726	10.61436
R(22)	942.9947	10.60451	10.60162
R(24)	944.1040	10.59205	10.58916
R(26)	945.1901	10.57988	10.57699
R(28)	946.2530	10.56799	10.56511
R(30)	947.2925	10.55640	10.55352
R(32)	948.3085	10.54509	10.54221
R(34)	949.3009	10.53406	10.53119
R(36)	950.2696	10.52332	10.52046
R(38)	951.2145	10.51287	10.51001
R(40)	952.1354	10.50270	10.49984
R(42)	953.0200	10.49295	10.49009
R(42)	953.0323	10.49282	10.48996
R(44)	953.9050	10.48322	10.48036
R(46)	954.7533	10.47390	10.47105
R(48)	955.5771	10.46487	10.46202
R(50)	956.3762	10.45613	10.45328
R(52)	957.1506	10.44767	10.44482
R(54)	957.8999	10.43950	10.43665
R(56)	958.6241	10.43161	10.42877
R(58)	959.3230	10.42401	10.42117

[a] From Freed *et al.* (1974, Table X). Frequencies in megahertz given by Freed *et al.* are not included here. Dry air wavelengths were calculated from the above data by this author.

TABLE A8

Calculated Wavenumbers and Wavelengths of the $^{13}C^{18}O_2$ Laser 00°1–[10°0, 02°0]$_{II}$ Band[a]

LINE	WV.N. (INV. CM.)	MICRONS(VAC)	MICRONS(AIR)
P(60)	976.1607	10.24421	10.24142
P(58)	978.1352	10.22353	10.22074
P(56)	980.0917	10.20312	10.20034
P(54)	982.0299	10.18298	10.18021
P(52)	983.9497	10.16312	10.16035
P(50)	985.8509	10.14352	10.14075
P(48)	987.7333	10.12419	10.12143
P(46)	989.5967	10.10512	10.10237
P(44)	991.4409	10.08632	10.08358
P(42)	993.2659	10.06779	10.06505
P(40)	995.0714	10.04952	10.04679
P(38)	996.8572	10.03152	10.02879
P(36)	998.6233	10.01378	10.01105
P(34)	1000.3695	9.99630	9.99358
P(32)	1002.0957	9.97908	9.97636
P(30)	1003.8017	9.96212	9.95941
P(28)	1005.4874	9.94542	9.94271
P(26)	1007.1528	9.92897	9.92627
P(24)	1008.7976	9.91279	9.91008
P(22)	1010.4219	9.89685	9.89415
P(20)	1212.0254	8.25065	8.24840
P(18)	1013.6082	9.86574	9.86305
P(16)	1015.1701	9.85056	9.84788
P(14)	1016.7110	9.83563	9.83295
P(12)	1018.2310	9.82095	9.81827
P(10)	1019.7299	9.80651	9.80384
P(8)	1021.2076	9.79232	9.78965
P(6)	1022.6642	9.77838	9.77571
P(4)	1024.0996	9.76467	9.76201
P(2)	1025.5137	9.75121	9.74855
P(0)	1026.9065	9.73798	9.73533
V(0)	1027.5949	9.73146	9.72880
R(2)	1028.9557	9.71859	9.71594
R(4)	1030.2952	9.70595	9.70331
R(6)	1031.6133	9.69355	9.69091
R(8)	1032.9101	9.68138	9.67874
R(10)	1034.1856	9.66944	9.66680
R(12)	1035.4397	9.65773	9.65510
R(14)	1036.6725	9.64624	9.64361
R(16)	1037.8840	9.63498	9.63236
R(18)	1039.0742	9.62395	9.62132
R(20)	1040.2432	9.61313	9.61051
R(22)	1041.3910	9.60254	9.59992
R(24)	1042.5176	9.5921635	9.5895492
R(26)	1043.6232	9.5820021	9.5793906
R(28)	1044.7077	9.5720549	9.5694461
R(30)	1045.7713	9.5623199	9.5597138
R(32)	1046.8139	9.5527954	9.5501919
R(34)	1047.8358	9.5434794	9.5408784
R(36)	1048.8370	9.5343699	9.5317714
R(38)	1049.8175	9.5254650	9.5228689
R(40)	1050.7774	9.5167627	9.5141690
R(42)	1051.7170	9.5082610	9.5056696
R(44)	1052.6362	9.4999578	9.4973686
R(46)	1053.5352	9.4918511	9.4892641
R(48)	1054.4142	9.4839388	9.4813540
R(50)	1055.2732	9.4762188	9.4736361
R(52)	1056.1124	9.4686890	9.4661083
R(54)	1056.9319	9.4613472	9.4587685
R(56)	1057.7319	9.4541912	9.4516145
R(58)	1058.5125	9.4472189	9.4446441

[a] From Freed *et al.* (1974, Table XI). Frequencies in megahertz given by Freed *et al.* are not included here. Dry air wavelengths were calculated from the above data by this author.

TABLE A9

Wavenumbers and Wavelengths for the CO Laser[a]

LINE	WV.N. (INV. CM.)	MICRONS(VAC)	MICRONS(AIR)
P 3(13)	2012.737	4.968359	4.967005
P 3(14)	2008.553	4.978709	4.977351
P 3(15)	2004.339	4.989176	4.987816
P 4(10)	1999.148	5.002131	5.000767
P 4(11)	1995.108	5.012260	5.010894
P 4(12)	1991.029	5.022529	5.021159
P 4(13)	1986.917	5.032923	5.031551
P 4(14)	1986.769	5.033298	5.031926
P 4(15)	1978.589	5.054107	5.052729
P 4(16)	1974.376	5.064891	5.063511
P 4(17)	1970.136	5.075792	5.074408
P 4(18)	1965.835	5.086897	5.085510
P 5(10)	1973.298	5.067658	5.066277
P 5(11)	1969.288	5.077977	5.076593
P 5(12)	1965.241	5.088434	5.087047
P 5(13)	1961.163	5.099015	5.097625
P 5(14)	1957.051	5.109729	5.108336
P 5(15)	1952.908	5.120569	5.119173
P 5(16)	1948.727	5.131555	5.130156
P 5(17)	1944.527	5.142639	5.141237
P 5(18)	1940.251*	5.153972	5.152567
P 5(19)	1936.002*	5.165284	5.163876
P 6(9)	1951.455	5.124382	5.122985
P 6(10)	1947.514	5.134751	5.133351
P 6(11)	1943.530	5.145277	5.143874
P 6(12)	1939.530	5.155888	5.154483
P 6(13)	1935.484	5.166666	5.165258
P 6(14)	1931.410	5.177565	5.176153
P 6(15)	1927.300	5.188606	5.187191
P 6(16)	1923.154†	5.199792	5.198374
P 6(17)	1918.986	5.211085	5.209665
P 7(9)	1925.711	5.192887	5.191472
P 7(10)	1921.808	5.203433	5.202015
P 7(11)	1917.864	5.214134	5.212713
P 7(12)	1913.894	5.224950	5.223525
P 7(13)	1909.880	5.235931	5.234504
P 7(14)	1905.846	5.247014	5.245583
P 7(15)	1901.766	5.258265	5.256831
P 7(16)	1897.663	5.269640	5.268203
P 7(17)	1893.518*	5.281175	5.279735
P 7(18)	1889.365*	5.292784	5.291341
P 8(8)	1903.877	5.252440	5.251008
P 8(9)	1900.043	5.263039	5.261604
P 8(10)	1896.174	5.273778	5.272340
P 8(11)	1892.262	5.284680	5.283240
P 8(12)	1888.326	5.295696	5.294252
P 8(13)	1884.352	5.306864	5.305417
P 8(14)	1880.340	5.318187	5.316737
P 8(15)	1876.305	5.329624	5.328171
P 8(16)	1872.231	5.341221	5.339765
P 8(17)	1868.118*	5.352981	5.351522
P 8(18)	1864.002*	5.364801	5.363339

[a] Wavenumbers in Table A9 were taken from Mantz *et al.* (1970, Table I) and from Roh *et al.* (1974, Table I). Numbers denoted by an asterisk are observed wavenumbers from Mantz *et al.*; numbers denoted by a dagger are calculated wavenumbers from Roh *et al.*; numbers with no notation are observed wavenumbers from Roh *et al.* All vacuum and dry air wavelengths were calculated by this author.

[b] The number following P in the table denotes the v'' value of the transition.

TABLE A9 (*Continued*)

LINE	WV.N. (INV. CM.)	MICRONS(VAC)	MICRONS(AIR)
P 9(8)	1878.258	5.324082	5.322631
P 9(9)	1874.457	5.334878	5.333424
P 9(10)	1870.616	5.345833	5.344375
P 9(11)	1866.745	5.356918	5.355458
P 9(12)	1862.844	5.368136	5.366673
P 9(13)	1858.903	5.379517	5.378050
P 9(14)	1854.930	5.391039	5.389569
P 9(15)	1850.929	5.402692	5.401220
P 9(16)	1846.918	5.414426	5.412949
P 9(17)	1842.821 *	5.426463	5.424984
P10(8)	1852.711	5.397496	5.396024
P10(9)	1848.943	5.408496	5.407021
P10(10)	1845.137	5.419652	5.418174
P10(11)	1841.301	5.430943	5.429462
P10(12)	1837.432	5.442378	5.440895
P10(13)	1833.526	5.453972	5.452485
P10(14)	1829.590	5.465705	5.464216
P10(15)	1825.620	5.477591	5.476098
P10(16)	1821.623 *	5.489610	5.488114
P10(17)	1817.606 *	5.501742	5.500243
P11(8)	1827.240	5.472735	5.471243
P11(9)	1823.509	5.483932	5.482438
P11(10)	1819.739	5.495294	5.493796
P11(11)	1815.936	5.506802	5.505301
P11(12)	1812.101	5.518456	5.516952
P11(13)	1808.232	5.530264	5.528757
P11(14)	1804.333	5.542214	5.540704
P11(15)	1800.402	5.554315	5.552801
P11(16)	1796.449 *	5.566537	5.565020
P12(7)	1805.509	5.538604	5.537095
P12(8)	1801.857	5.549830	5.548318
P12(9)	1798.156	5.561253	5.559737
P12(10)	1794.419	5.572834	5.571316
P12(11)	1790.653	5.584555	5.583033
P12(12)	1786.853	5.596431	5.594906
P12(13)	1783.018	5.608468	5.606940
P12(14)	1779.162	5.620624	5.619092
P12(15)	1775.255	5.632994	5.631458
P13(7)	1780.179	5.617413	5.615882
P13(8)	1776.550	5.628887	5.627353
P13(9)	1772.891 †	5.640505	5.638968
P13(10)	1769.191	5.652301	5.650761
P13(11)	1765.459	5.664249	5.662706
P13(12)	1761.696 †	5.676348	5.674801
P13(13)	1757.896	5.688619	5.687068
P13(14)	1754.064	5.701046	5.699493
P13(15)	1750.204	5.713620	5.712063
P14(6)	1758.494	5.686684	5.685134
P14(7)	1754.932	5.698227	5.696674
P14(8)	1751.335 †	5.709930	5.708374
P14(9)	1747.707	5.721783	5.720224
P14(10)	1744.041	5.733810	5.732248
P14(11)	1740.347	5.745981	5.744415
P14(12)	1736.611	5.758342	5.756773
P14(13)	1732.855	5.770823	5.769251
P14(14)	1729.053	5.783513	5.781937

TABLE A9 (*Continued*)

LINE	WV.N. (INV. CM.)	MICRONS(VAC)	MICRONS(AIR)
P15(5)	1736.777	5.757792	5.756223
P15(6)	1733.288†	5.769382	5.767809
P15(7)	1729.761†	5.781146	5.779570
P15(8)	1726.198	5.793078	5.791499
P15(9)	1722.603	5.805168	5.803586
P15(10)	1718.981	5.817400	5.815815
P15(11)	1715.312	5.829843	5.828254
P15(12)	1711.618	5.842425	5.840833
P15(13)	1707.887	5.855188	5.853593
P16(6)	1708.166	5.854232	5.852637
P16(7)	1704.677	5.866214	5.864615
P16(8)	1701.149†	5.878380	5.876778
P16(9)	1697.592	5.890697	5.889092
P16(10)	1693.994	5.903209	5.901600
P16(11)	1690.371	5.915861	5.914249
P16(12)	1686.703	5.928726	5.927110
P16(13)	1683.017†	5.941711	5.940091
P16(14)	1679.289	5.954901	5.953278
P17(7)	1679.674	5.953536	5.951914
P17(8)	1676.183	5.965936	5.964310
P17(9)	1672.656	5.978516	5.976886
P17(10)	1669.100†	5.991253	5.989620
P17(11)	1665.508	6.004174	6.002538
P17(12)	1661.881	6.017278	6.015638
P17(13)	1658.212	6.030592	6.028949
P18(8)	1651.303	6.055824	6.054173
P18(9)	1647.816	6.068639	6.066985
P18(10)	1644.294	6.081638	6.079980
P18(11)	1640.733	6.094837	6.093176
P18(12)	1637.136	6.108228	6.106563
P18(13)	1633.530	6.121712	6.120044
P19(6)	1633.320	6.122499	6.120830
P19(7)	1629.929	6.135237	6.133565
P19(8)	1626.508	6.148141	6.146465
P19(9)	1623.056	6.161217	6.159538
P19(10)	1619.568	6.174486	6.172803
P19(11)	1616.043	6.187954	6.186268
P19(12)	1912.491	5.228783	5.227357
P19(13)	1608.909	6.215392	6.213698
P20(6)	1608.531	6.216853	6.215158
P20(7)	1605.180	6.229831	6.228133
P20(8)	1601.795	6.242996	6.241295
P20(9)	1598.376	6.256350	6.254645
P20(10)	1594.921	6.269903	6.268194
P20(11)	1591.435	6.283637	6.281925
P20(12)	1587.917	6.297558	6.295842
P21(5)	1587.106	6.300776	6.299059
P21(6)	1583.831	6.313805	6.312084
P21(7)	1580.516	6.327048	6.325323
P21(8)	1577.160	6.340511	6.338783
P21(9)	1573.775	6.354149	6.352417
P21(10)	1570.355	6.367987	6.366252
P21(11)	1566.902	6.382020	6.380281

TABLE A9 *(Continued)*

LINE	WV.N. (INV. CM.)	MICRONS(VAC)	MICRONS(AIR)
P22(5)	1562.459	6.400168	6.398424
P22(6)	1559.210 †	6.413504	6.411757
P22(7)	1555.929	6.427029	6.425277
P22(8)	1552.611	6.440763	6.439008
P22(9)	1549.263	6.454682	6.452923
P22(10)	1545.881	6.468803	6.467040
P23(5)	1537.886	6.502433	6.500661
P23(6)	1534.669	6.516063	6.514287
P23(7)	1531.420	6.529887	6.528108
P23(8)	1528.139	6.543907	6.542124
P23(9)	1524.822	6.558143	6.556355
P23(10)	1521.479 †	6.572552	6.570761
P23(11)	1518.097	6.587194	6.585399
P23(12)	1514.679	6.602059	6.600260
P23(13)	1511.230	6.617126	6.615323
P24(5)	1513.388	6.607691	6.605890
P24(6)	1510.206	6.621613	6.619809
P24(7)	1506.997 †	6.635713	6.633905
P24(8)	1503.748	6.650050	6.648238
P24(9)	1500.474	6.664561	6.662744
P24(10)	1497.156	6.679331	6.677511
P24(11)	1493.812	6.694283	6.692459
P24(12)	1490.430	6.709473	6.707645
P25(4)	1492.080	6.702054	6.700227
P25(5)	1488.967 †	6.716066	6.714235
P25(6)	1485.824	6.730272	6.728438
P25(7)	1482.646	6.744698	6.742860
P25(8)	1479.439	6.759319	6.757477
P25(9)	1476.189 †	6.774200	6.772354
P25(10)	1472.910	6.789281	6.787431
P25(11)	1469.597	6.804587	6.802732
P25(12)	1466.252	6.820110	6.818251
P25(13)	1462.874	6.835859	6.833996
P26(6)	1461.510	6.842239	6.840374
P26(7)	1458.366 †	6.856989	6.855121
P26(8)	1455.189 †	6.871960	6.870087
P26(9)	1451.991	6.887095	6.885218
P26(10)	1448.740	6.902550	6.900669
P26(11)	1445.458	6.918222	6.916337
P26(12)	1442.141	6.934135	6.932245
P27(5)	1440.341	6.942800	6.940908
P27(6)	1437.263 †	6.957669	6.955773
P27(7)	1434.155	6.972747	6.970847
P27(8)	1431.015	6.988047	6.986143
P27(9)	1427.842	7.003576	7.001667
P27(10)	1424.630	7.019366	7.017454
P27(11)	1421.385	7.035392	7.033474
P28(6)	1413.084	7.076720	7.074792
P28(7)	1410.009	7.092153	7.090221
P28(8)	1406.901	7.107821	7.105884
P28(9)	1403.754	7.123755	7.121814
P28(10)	1400.586	7.139869	7.137923
P28(11)	1397.382	7.156239	7.154289

TABLE A9 *(Continued)*

LINE	WV.N. (INV. CM.)	MICRONS(VAC)	MICRONS(AIR)
P29(6)	1388.954	7.199663	7.197701
P29(7)	1385.919	7.215429	7.213463
P29(8)	1382.851	7.231437	7.229466
P29(9)	1379.740	7.247742	7.245767
P29(10)	1376.595	7.264301	7.262321
P30(6)	1364.889	7.326603	7.324607
P30(7)	1361.888	7.342748	7.340747
P30(8)	1358.855	7.359137	7.357131
P30(9)	1355.779	7.375833	7.373823
P30(10)	1352.674	7.392764	7.390750
P31(6)	1340.877	7.457806	7.455773
P31(7)	1337.903	7.474383	7.472347
P31(8)	1334.906	7.491164	7.489123
P31(9)	1331.872	7.508229	7.506183
P31(10)	1328.796	7.525610	7.523559
P31(11)	1325.697	7.543202	7.541146
P32(7)	1313.967	7.610541	7.608467
P32(8)	1311.002	7.627753	7.625675
P32(9)	1307.998	7.645272	7.643188
P32(10)	1304.965	7.663041	7.660953
P32(11)	1301.891	7.681135	7.679041
P33(6)	1292.958	7.734203	7.732096
P33(7)	1290.066	7.751542	7.749429
P33(8)	1287.131	7.769217	7.767100
P33(9)	1284.166	7.787155	7.785033
P33(10)	1281.161	7.805420	7.803293
P33(11)	1278.132	7.823918	7.821786
P34(6)	1269.050	7.879910	7.877763
P34(7)	1266.183	7.897753	7.895600
P34(8)	1263.288	7.915851	7.913694
P34(9)	1260.356 †	7.934266	7.932104
P34(10)	1257.385	7.953014	7.950846
P35(6)	1245.167	8.031051	8.028863
P35(7)	1242.330	8.049391	8.047198
P35(8)	1239.467	8.067984	8.065785
P35(9)	1236.573	8.086866	8.084662
P35(10)	1233.642	8.106079	8.103870
P36(7)	1218.489	8.206886	8.204649
P36(8)	1215.664	8.225957	8.223716
P36(9)	1212.782	8.245505	8.243258
P36(10)	1209.892	8.265200	8.262948

References

Chang, T. Y. (1970). *Opt. Commun.* **2**, 77.

Claspy, P. C., Pao, Y. -H., Kwong, S., and Nodov, E. (1976). *Appl. Opt.* **15**, 1506.

Coleman, C. D., Bozman, W. R., and Meggers, W. F. (1960). Table of Wavenumbers. Natl. Bur. Std. Monograph 3.

Cottrell, T. L., Macfarlane, I. M., Read, A. W., and Young, A. H. (1966). *Trans. Faraday Soc.* **62**, 2655.

Dailey, W. V. (1967). *Proc. Ann. Anal. Instrum. Symp. 11th*, p. 189. Plenum Press, New York.

Dewey, C. F. Jr. (1974). *Opt. Eng.* **13**, 483.

Dewey, C. F., Jr., Kamm, R. D., and Hackett, C. E. (1973). *Appl Phys. Lett.* **23**, 633.

Freed, C. (1970). *Int. Electron Devices Meeting, Washington, D.C.* Paper 14.4.

Freed, C. (1971). *Appl. Phys. Lett.* **18**, 458.

Freed, C., Ross, A. H. M., and O'Donnell, R. G. (1974). *J. Mol. Spectrosc.* **49**, 439.

Gelbwachs, J. (1974). *Appl. Opt.* **13**, 1005.

Goldan, P. D., and Goto, K. (1974). *J. Appl. Phys.* **45**, 4350.

Goldberg, M. W., and Yusek, R. (1970). *Appl. Phys. Lett.* **17**, 349.

Hill, D. W., and Powell, T. (1968). "Non-Dispersive Infra-Red Gas Analysis in Science, Medicine and Industry." Plenum Press, New York.

Kaiser, R. (1959). *Can. J. Phys.* **37**, 1499.

Kerr, E. L., and Atwood, J. G. (1968). *Appl. Opt.* **7**, 915.

Knox, J. D., and Pao, Y. -H. (1971). *Appl. Phys. Lett.* **18**, 360.

Kreuzer, L. B. (1971). *J. Appl. Phys.* **42**, 2934.

Kreuzer, L. B. (1974). *Anal. Chem.* **46**, 235A.

Kreuzer, L. B., and Patel, C. K. N. (1971). *Science* **173**, 45.

Kreuzer, L. B., Kenyon, N. D., and Patel, C. K. N. (1972). *Science* **177**, 347.

Luft, K. F. (1943). *Z. Tech. Phys.* **24**, 97.

Mantz, A. W., Nichols, E. R., Alpert, B. D., and Rao, K. N. (1970). *J. Mol. Spectrosc.* **35**, 325.

Microphones and Microphone Preamplifiers (1975). Brüel and Kjair, DK-2850, Naerum, Denmark.

Occupational Safety and Health Administration Concentration Limits for Gases (1972). *Fed. Register* **37**, No. 22140, October 18.

Pao, Y. -H., and Claspy, P. C. (1975). An investigation of the feasibility of use of laser opto-acoustic detection for the detection of explosives. Case Western Reserve Univ., Final Rep. for Subcontract 44343-V, The Aerospace Corp.

Patel, C. K. N., Burkhardt, E. G., and Lambert, C. A. (1974). *Science* **184**, 1173.

Pristera, F., Halik, M., Castelli, A., and Fredericks, W. (1960). *Anal. Chem.* **32**, 495.

Roh, W. B., and Rao, K. N. (1974). *J. Mol. Spectrosc.* **49**, 317.

Slobodskaya, P. V. (1948). *Izv. Akad. Nauk SSSR Ser. Fiz.* **12**, 656.

Veingerov, M. L. (1938). *Dokl. Akad. Nauk SSSR* **19**, 687.

7

Photoacoustic Spectroscopy of Gases in the Visible and Ultraviolet Spectral Regions

M. B. Robin

Bell Laboratories
Murray Hill, New Jersey

I. Introduction

The demonstration by Kreuzer and Patel at Bell Laboratories of the high sensitivity of the photoacoustic technique in the infrared region (Kreuzer and Patel, 1971; Kreuzer, 1971) suggested immediately that the same technique in the visible–ultraviolet region could be of great interest to both spectroscopists and photochemists. That such a technique could yield signals of measurable size was already demonstrated much earlier by Bell

(1881), who detected the sound waves aurally from colored materials illuminated periodically with white light.

To date, photoacoustic spectroscopy in the infrared and visible–ultraviolet regimes have developed along somewhat different lines. In the infrared, interest in photoacoustic spectroscopy has centered about the analytical determination of trace-gas contaminants using lasers of fixed frequency for excitation, and also about the determination of the vibrational lifetimes of very simple molecules excited into their lower vibrational levels. In each case, the experiment is performed at a fixed wavelength, appropriate to the molecular vibration. In contrast, the use of photoacoustic spectroscopy in the visible–ultraviolet region for analytical purposes is virtually unexplored, and furthermore, the question of obtaining relaxation times for levels excited in this region is far more complicated than that for infrared absorption.

In the case of electronic excitation, one is dealing with an initial excitation corresponding energetically to perhaps 20 quanta of vibration, and the pathway to heat is far less certain. Hence, the meaning of a lifetime measured at one wavelength, as done in the infrared, is less clear for an electronic transition (though such studies have been done). As stressed by Cottrell *et al.* (1966), the relaxation time (lifetime) measured photoacoustically is not necessarily that appropriate to the level initially excited, but may correspond instead to the average lifetime of energy being locked up in a variety of modes, before becoming heat and activating the microphone. Though originally enunciated for vibrational excitation, this complication will be especially important for electronic excitations in large molecules, where the intermediate levels may be either vibrations or low-lying electronic states of the molecule. It may happen as well that the intermediate levels are instead within other molecules (purposely or accidentally present), in which case, the kinetics of the heat release may be complicated by both the rates of energy transfer to and release from the intermediate level.

In addition to the orders-of-magnitude higher complexity of the relaxation from an energetic electronic excitation, photoacoustic spectroscopy in the infrared is rarely done in the scanning mode, whereas with incoherent lamps or scanning dye lasers of high power and high spectral resolution, such an approach is extremely attractive in the visible–ultraviolet region. The extreme complexity of the relaxation from a level 30,000 cm^{-1} above the ground state, with several intermediate electronic states (each with its own ladder of vibrational–rotational sublevels) is mind boggling, but one hopes that such scanning photoacoustic experiments can reveal the thresholds of some of the more obvious intermediate steps in the overall relaxation.

For reasons of heightened sensitivity, it seems certain that eventually laser-excited photoacoustic spectroscopy in the visible–ultraviolet region will find analytical uses. First, the oscillator strengths for allowed electronic excitations are more than ten times larger than for vibrational excitations, and second, the photon energy, $h\nu$, to be degraded into heat is about ten times larger in the visible than in the infrared. Still, in spite of these advantages, virtually no work has been done in this area (however, see the discussion on NO_2). Perhaps this reluctance is a result of the fact that all substances will show infrared or Raman absorption whereas many molecules cannot be excited electronically in the visible–ultraviolet region.*

Upon excitation into an electronically excited state, a molecule has three options open to it:

1. It may luminesce to a vibrational level of the ground state which then relaxes further to produce a relatively small amount of heat in its surroundings.
2. It may undergo photochemistry to produce new chemical species with the concomitant deposition of more or less heat in the surroundings.
3. It may "do nothing," i.e., cascade down the myriad ladders of the various electronic states to finally reach the ground state, with the conversion of the full $h\nu$ into heat.

Inasmuch as these various pathways compete with one another to extents depending upon their relative rates ($\Phi_{\text{lum}} + \Phi_{\text{photochem}} + \Phi_{\text{nr}} = 1$, where the three quantum yields are those for luminescence, photochemistry, and nonradiative transitions, respectively), one sees that the direct study of the heat generated following excitation is of immediate consequence to luminescence and photochemical studies, in addition to radiationless transitions. This is illustrated further in Section IV.

The photoacoustic spectrometer may be scanned advantageously in either of two general modes:

1. recording amplitude information, relying upon the sensitivity of the technique to give an absorptionlike spectrum for systems of low optical density, and
2. recording phase information which reflects the overall rate of conversion of ro-vibronic excitation energy into translational energy of the surrounding gas.

In this second mode, one can profitably measure aspects of the relaxation kinetic scheme absolutely at one wavelength, or determine the relative relaxation kinetics as the absorption spectrum is scanned. Alternatively, kinetic information can be obtained by using pulsed optical excitation, and determining the pressure pulse shape by using a boxcar integrator or signal

* Actually, photoacoustic signals have been observed in "colorless" gases such as NH_3 and CH_4 when placed in the cavity of a visible dye laser (Stella *et al.*, 1975). The absorption is to the overtone bands of the rotational–vibrational manifold, as in thermal-blooming spectroscopy in the visible region (Whinnery, 1974).

averager. If excitations at all points of a spectrum relax at the same rate and with a constant quantum yield of heat, then the heat-amplitude spectrum will resemble the absorption spectrum, the only difference owing to the fact that for equal absorption intensities at two points, the higher-frequency point will yield more heat. Such heat-amplitude spectra will be grossly distorted if the rates or quantum yields change in the spectral region under investigation (see Section IV).

II. Phase Angle and Amplitude

The key feature of the photoacoustic effect is that the application of radiation which is sine-wave modulated in intensity will result in pressure variations which are also sine-wave modulated and of the same frequency. However, on a common time scale, the time of maximum light intensity does not coincide with that of maximum pressure, for obviously, the molecules that are excited by the radiation must live their nonradiative lifetime before they can release their heat. Thus, the finite nonradiative lifetime leads to a phase difference between the incoming optical wave train and the outgoing acoustic wave train. All of our kinetic measurements revolve about the measurement of this phase lag.

The original mathematical analysis of the photoacoustic effect was aimed specifically at describing the situation wherein a vibration is pumped harmonically at frequency ω by infrared radiation (Read, 1967–1968; Delaney, 1959) and decays without populating intermediate levels. As Delaney (1959) describes it, once the light is turned on:

> the thermodynamic state of the gas was represented by two absolute temperatures, corresponding to two different groups of degrees of freedom of the gas molecules; one group consisting of those degrees (vibrational) which absorb the radiation directly and the other of the translational degrees. It was assumed that individually the two groups were in statistical equilibrium, although not necessarily mutually in thermal equilibrium, and the energy exchange between them was characterized by a relaxation time.

Because the analysis of such a situation in no way is unique for vibrational excitation, one can generalize to the case where the upper level of the two-level system is an electronic state instead.

Considering the respective heat capacities and the transfer of energy from the bath of electronic excitation to the thermal bath (which is assumed rapid with respect to the transfer of energy to the apparatus walls), and realizing that the rate at which the two temperatures equilibrate

$$d\,\Delta T/dt = -\Delta T/\tau, \tag{1}$$

then

$$\tan(\pi/2 - \phi) = \omega\tau, \tag{2}$$

where τ is the characteristic lifetime of the electronically excited species, and ϕ is the phase angle resulting from the nonzero lifetime of the excited state.* Similarly, one calculates that the maximum temperature fluctuation will be

$$\Delta T_{\max} = A(1 + \omega^2\tau^2)^{-1/2}/\omega C_0, \tag{3}$$

where A is a constant relating both to the incident light intensity and the oscillator strength of the transition in question and C_0 is the heat capacity of the gas. Of course, the photoacoustic amplitude will be proportional to $\Delta T_{\max}$. If the relaxation is initiated by collisions in the gas phase at pressure P, then the right-hand side of Eq. (2) is replaced by $\omega\tau/P$. These equations are appropriate for sinusoidally chopped excitation, and cannot be used with pulsed-laser excitation.

From Eqs. (1)–(3) one sees that at modulation frequencies much less than the relaxation rate ($\omega \ll \tau^{-1}$), the signal amplitude is inversely proportional to the frequency, and that ϕ must be close to $\pi/2$. The ω^{-1} dependence in Eq. (3) arises from the fact that the amount of light illuminating the system per cycle decreases as the chopping frequency increases. At higher frequencies where ω and τ^{-1} are approximately equal, the phase angle will then vary with chopping frequency as in Eq. (2), while at high frequencies, the amplitude will vary as ω^{-2}. At very high frequencies compared with τ^{-1}, few molecules with such long lifetimes will have an opportunity to relax within the time ω^{-1}, and so the amplitude will go to zero. In any event, it is important to pick ω and the cell dimensions such that a long τ will not lead to relaxation at the walls.

According to Cottrell and McCoubrey (1961), the photoacoustic pressure is directly proportional to the gas pressure for a gas of constant composition, and to the concentration of absorbing gas in mixtures of constant total pressure, density and specific heat.

A more realistic and detailed relaxation scheme for electronic excitation has been constructed by Hunter *et al.* (1974). Referring to Fig. 1, they consider harmonic optical excitation from S_0 to S_1 followed by simultaneous return to S_0 via both fluorescence (releasing heat E_1 with rate constant k_1) and non-radiative transitions (releasing heat E_2 with rate constant k_2), and inter-system crossing to the lowest triplet T_1 (releasing heat E_3 with rate constant k_3). Those molecules reaching T_1 then relax radiationlessly to release heat E_4

* If there are no intermediate levels which can act as traps for the heat, then the τ determined photoacoustically from Eq. (2) will equal that found experimentally from the luminescence decay curve. See the discussion in Section IV for examples in which this is the case (biacetyl, acetophenone), and one situation in which it does not apply (glyoxal).

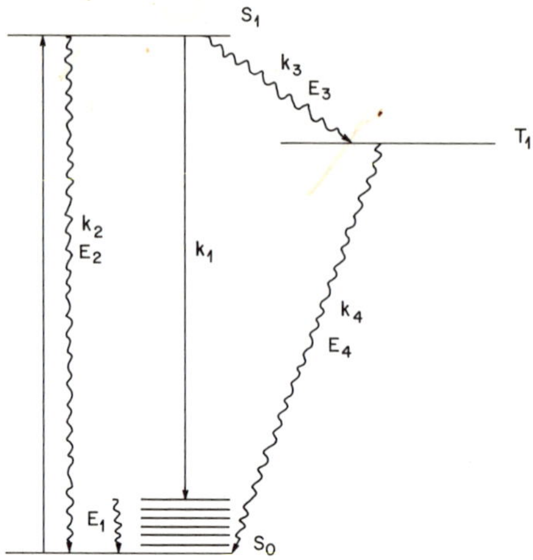

Fig. 1. Relaxation scheme for a typical organic molecule. (Adapted from Hunter and Stock, 1974.)

at rate k_4. Such a relaxation scheme is quite common among organic molecules. Considering the rates of population of the levels as the light is modulated, they show that for unimolecular decay at each step, the phase angle is given by

$$\tan\phi = L(\sin\theta_T \cos\theta_T)(H + L\cos^2\theta_T) \tag{4}$$

where $\cos\theta_T = (1 + \omega^2\tau_4{}^2)^{-1/2}$, provided the lifetimes in the singlet manifold are very much smaller than either τ_4 or ω^{-1}, which is almost always the case. In Eq. (4), L represents the amount of heat resulting from relaxation in the slow step and H is the sum of the heats which are much faster and either precede or parallel the excitation of T_1. Similarly, the amplitude is given by

$$P = (bF_0/C_0\omega)\{H^2 + \cos^2\theta_T(L^2 + 2HL\}^{1/2} \tag{5}$$

in which bF_0 is a measure of the absorbed power per unit volume. In this formulation, a plot of the phase angle ϕ versus ω has a maximum given by

$$\tan\phi_{max} = \tfrac{1}{2}\{(H/L)(H/L + 1\}^{-1/2}. \tag{6}$$

If the excitation is to S_2, followed by $S_2 \rightsquigarrow S_1$ radiationless relaxation, then this energy increment is simply added to the definition of H.

With these equations and the experimental determination of the dependence of the phase angle and amplitude upon chopping frequency, one can then obtain values for τ_4 and H/L. Characteristic amplitude and phase angle variations with chopping frequency are shown in Fig. 2 for a few values of H/L and τ_4.

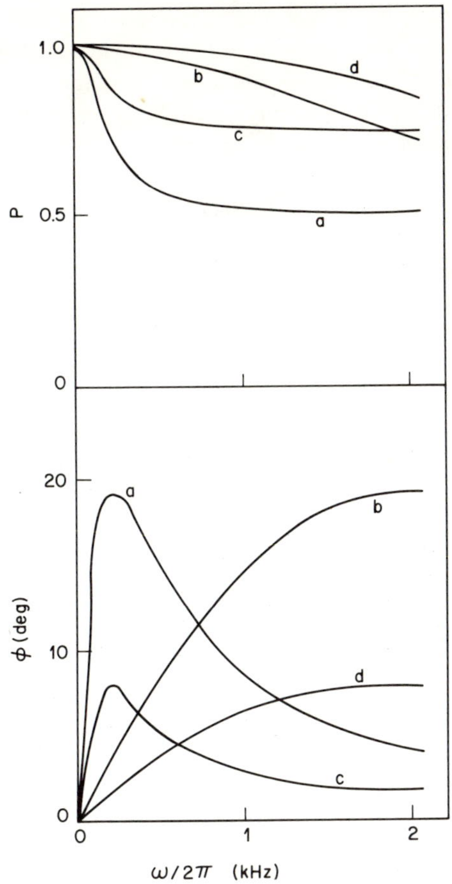

Fig. 2. Computed heat amplitudes (arbitrary units) and phase angles for various values of H/L and τ; (a) $\tau = 10^{-3}$ sec, $H/L = 1.0$; (b) $\tau = 10^{-4}$ sec, $H/L = 1.0$; (c) $\tau = 10^{-3}$ sec, $H/L = 3.0$; (d) $\tau = 10^{-4}$ sec, $H/L = 3.0$ (From Hunter and Stock, 1974a.)

If $L \gg H$ as would occur when E_3 is small and k_3 is very large so that virtually all of the molecules pass through T_1, then Eqs. (4) and (5) reduce to Eqs. (2) and (3), as given in the simpler two-level thermodynamic treatment.

The successful application of the above equations depends upon avoiding several experimental pitfalls which fatally marred earlier work in this field. First, heat from the relaxation can diffuse to the cell walls and introduce a frequency-dependent phase lead while decreasing the amplitude of the signal. This effect can be important at low pressures (below about 1 torr) and for chopping frequencies that are not much greater than the reciprocal of the characteristic time for heat conduction to the wall. In order to maximize photoacoustic sensitivity, it is advantageous to operate at or near a cell resonance, however, for meaningful measurements of the phase angle, one must be as far away from such resonances as possible. As mentioned above, the phase angle will also be strongly perturbed by any Helmholz resonances in the cell. Finally, infrared studies have shown that impurities at the parts per million level can be very active in promoting vibrational relaxation, and this should be considered as potentially important as well for electronic excitation (especially traces of oxygen). These and other mechanical complications to phase-angle measurement need not detour the experimentalist who would scan the wavelength at constant pressure and chopping frequency, and so get relative phase angles across the spectrum.

Measurement of the phase angle at any one wavelength for a polyatomic molecule will yield a single value, and hence a single value of the lifetime, even though there are in fact several consecutive and/or simultaneous processes of different lifetime participating in the overall relaxation. Consider each relaxation step as a vector, with magnitude proportional to the amount of heat released, and the tangent of its angle depending upon the lifetime [Eq. (2)]. In a complex manifold, imagine then a nonunique relaxation to the ground state (Path A), passing through successive steps, each with lifetime τ_i. In that case, the overall phase angle which one would measure (Φ_α) will be related to the component τ_i by

$$\tan \Phi_\alpha = \sum_i \tan \phi_i. \tag{7}$$

With several parallel paths to the ground state, each with its own overall phase angle Φ_α and heat amplitude A_α, the overall heat vector is then the vector sum of these competing-process vectors, and

$$\tan \Phi_\mathrm{T} = \sum_\alpha A_\alpha \sin \Phi_\alpha \Big/ \sum_\alpha A_\alpha \cos \Phi_\alpha. \tag{8}$$

Equation (8) is obviously complicated, but if only one path is important to the ground state, and if in that chain only one ϕ_i is large, then the measurement of Φ_T will give the lifetime of this ith level via Eq. (2). In the more general case, the lifetime measured directly from Φ_T via Eq. (2) has no meaning in terms of the lifetime of a particular level. In this regard, one can do somewhat better using an analysis of the heat-pulse shape, which displays the various component heats directly on the time axis.

III. Experimental Technique

In this section, the experimental apparatus is briefly described which we have found most suitable for photoacoustic spectroscopy in the visible–ultraviolet region. For illumination, a pulsed dye laser is ideal optically, but the pulse rate cannot go beyond 1000 Hz, and is much lower in most. The restriction on pulse frequency can be overcome using a CW dye laser and external chopping. Lacking such devices, a high-pressure xenon arc lamp of 2–4-kW output dispersed by a 0.25-m monochromator can be used. With 1-mm slits, such an arrangement yields about 0.5 mW in a spectral band of $\sim 1000\,\text{cm}^{-1}$ at any point between 2000 and 10,000 Å, the visible–ultraviolet region. This light is chopped with either a variable-speed spinning slotted disk or a tuning-fork chopper (preferred) in the range 30–3000 Hz, and the light pulses focused into a steel cell containing a foil-electret microphone either of the cylindrical type used by Kreuzer and Patel (1971, Kreuzer, 1971) or the disk type described by Sessler and West (1964, 1966). We prefer the latter type, with the microphone secreted in the sidearm of a T-shaped cell. At the rear of the cell, a bellows-mounted mirror reflects unabsorbed light back into the cell, thus doubling the signal for weak absorbances, and it can be translated so as to tune the cell to an acoustic resonance. The light should not be allowed to strike either the walls or the microphone itself; in this sense, photoacoustic spectroscopy is not unaffected by scattered light. The sensitivity to be gained by using a tuned cell is readily demonstrated by the experimental results in Fig. 3, in which the microphone response is plotted against the chopping frequency for a mixture of ammonia in air. The signal peaks correspond to a tuned condition in one or more of the three acoustic eigenmodes of the chamber (Max and Rosengren, 1974). Quantitative aspects of photoacoustic cell design have recently been discussed by Rosengren (1975).

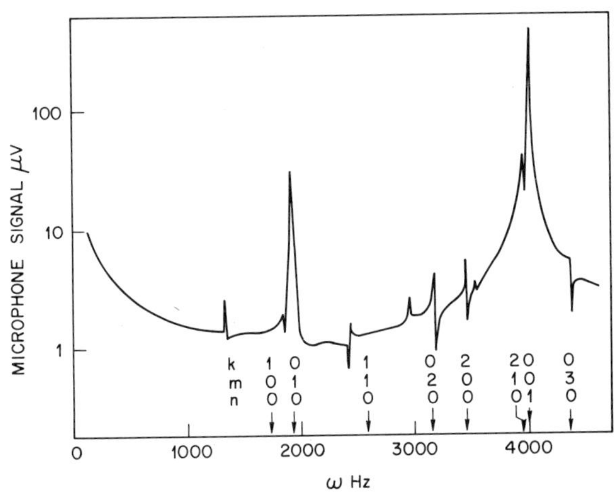

Fig. 3. Microphone response as a function of the chopping frequency, for absorption by 10 ppm of ammonia in air. The calculated cell resonances are also shown, together with the acoustical mode eigenvalues, where k, m, and n are the longitudinal, azimuthal, and radial mode quantum numbers, respectively. (From Max and Rosengren, 1974.)

Best results were obtained using an electret foil made from 0.0005-in. Mylar coated with 200 Å of aluminum; the foil was permanently charged before assembly of the microphone by holding a discharging Tesla coil above the film (metal side down) for 3–5 min without letting the discharge puncture the film.* The microphone signal is amplified $\times 100$ using a voltage-sensitive preamp (such as Ithaco model 144L), and the signal fed to either a lock-in detector or a vector voltmeter. In the case of highly reactive samples, an electret foil consisting of gold on Teflon has been used successfully. Externally biased capacitance microphones drift badly in chemically reactive environments.

The spectra so obtained can be normalized by dividing the spectra by that of carbon black, obtained either sequentially or simultaneously as in a double-beam arrangement using two cells and an achromatic beam splitter.

The microphone, being a mechanical device, is limited in its frequency response to $\sim 2 \times 10^5$ Hz, which is also the upper limit of the amplifier response. This places the upper limit on the relaxation rate one can measure at $\sim 10^6\ \text{sec}^{-1}$, while at the lower end, rates slower than $10^2\ \text{sec}^{-1}$ will be adversely affected by the flow of excited molecules to the walls and relaxation there (Read, 1967–1968). Heat amplitude can be readily observed for relaxations with rates faster than $10^6\ \text{sec}^{-1}$, but the rate cannot be measured.

One important experimental factor in the measurement of the lifetime from the phase angle is the strict avoidance of Helmholz resonances. To this end, dead spaces in the cell other than the illuminated volume must be avoided in the cell design, or filled with pellets to forestall the resonant transfer of the acoustic energy within the different parts of the cell. Though signals of millivolt amplitude have been reported to emanate from borosilicate windows when illuminated with white light (Parker, 1973), this has never been a factor when a monochromator was used in the visible–ultraviolet region.

As for sensitivity, Kreuzer and Patel (1971) report that with 7 mW of light incident on the cell at 3.39 μm, they could detect one part in 10^8 of methane in an atmosphere of N_2. The absorption coefficient of methane at this wavelength is 100. Thus, in a pure gas under the same conditions, the limit of detectability (signal/noise = 1) is an absorption having an absorption coefficient of about 10^{-6}, corresponding to an oscillator strength of $\sim 10^{-11}$. More refined calculations by Rosengren (1973) conclude that the sensitivity may be increased beyond this by a factor of $\times 100$, and this was confirmed using ammonia and a 1-W infrared laser (Max and Rosengren, 1974).

* We have not tried any of the commercial microphones, but Butcher *et al.* (1975) report success with a 2.5-cm General Radio Electret Microphone, and several others have used the more expensive Brüel and Kjair microphone.

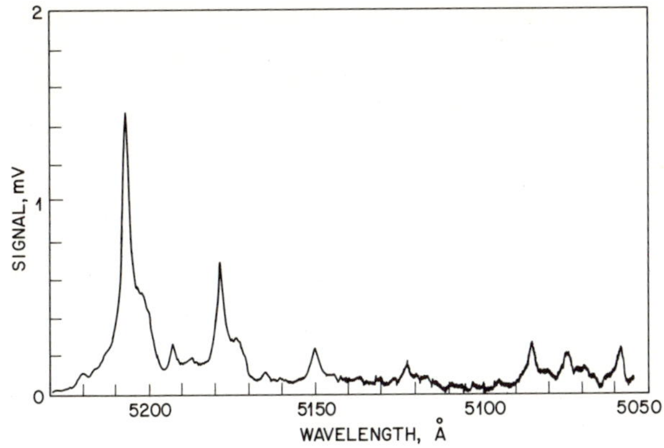

Fig. 4. The optoacoustic spectrum of the $S_0 \rightarrow T_1$ transition in glyoxal at 37 Hz, 50 torr, using a 4-cm path. (From Kaya *et al.*, 1975b.)

Such high sensitivity in the photoacoustic technique implies the ready observation of formally forbidden excitations such as singlet → triplet, etc. This sensitivity has not yet been exploited in the visible-ultraviolet region, but one example can be given. In Fig. 4, the $S_0 \rightarrow T_1$ absorption band of glyoxal vapor is shown, as recorded using 50-torr pressure in a 4-cm cell and dye laser excitation (Kaya *et al.*, 1975b). There is nothing new here spectroscopically, the spectrum having been observed by others in more conventional ways, but it is to be remembered that the others typically use path lengths of several meters for such spectra, compared to 0.04 m used here. The oscillator strength of this band in glyoxal has not been measured, but the corresponding band in biacetyl has an oscillator strength of 1.4×10^{-7}. Using this figure for glyoxal leads to the conclusion that the small features at 5190 Å, etc., have optical densities of 10^{-4}–10^{-5}. Stella *et al.* (1975) report remarkable sensitivity by placing the absorption cell within the laser resonator.

With modest excitation powers (0.5 mW) and moderate absorption coefficients ($\varepsilon \sim 3000$), we have obtained spectra at absorber pressures of ~ 10 mtorr. With certainty, signals can be obtained at even lower pressures, but in that case, relaxation at the cell walls must be considered. The amplitude of the heat signal from low-pressure absorbers is increased considerably by adding a second gas such as N_2, He, etc., but these may have profound effects on the relaxation kinetics.

Because the lock-in amplifier processes only that component of the total heat flux having the requisite frequency, it is relatively simple to illuminate a system with dc light and simultaneously probe the photoacoustic spectrum of the reaction intermediates with a second colinear modulated light beam; the same experiment with a photomultiplier tube is very difficult due to saturation of the detector.

IV. Illustrative Spectra

There is so little work of this type in the literature that one cannot present a very unified or coherent description of photoacoustic spectroscopy in the visible–ultraviolet region. Suffice it then, to present scattered instances, which may be taken to represent the various types of phenomena that can be encountered in photoacoustic spectroscopy.

A. Nitrogen Dioxide

As the first example, consider NO_2, the absorption spectrum of which is shown in Fig. 5 (Harshbarger and Robin, 1973a). Though taken with far lower resolution, the photoacoustic spectrum is seen to follow the absorption up

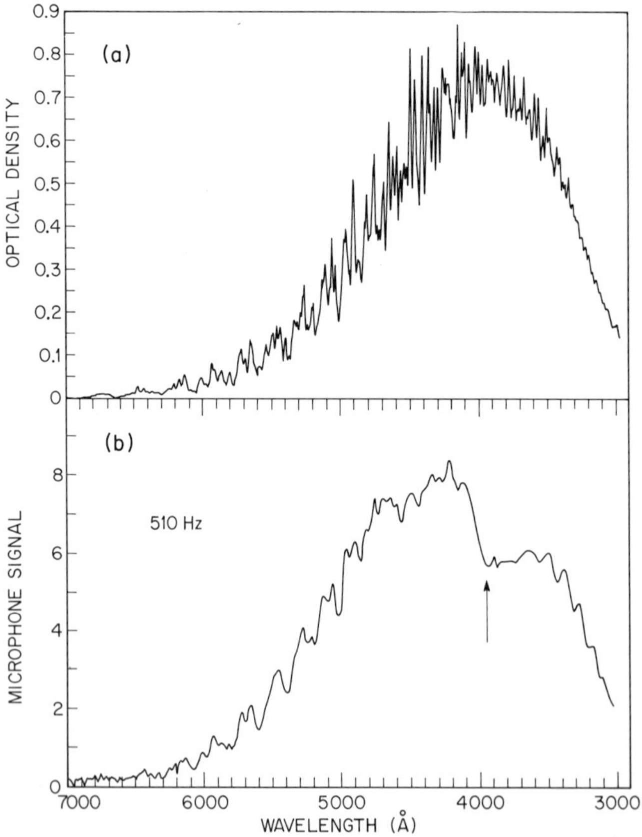

Fig. 5. Comparison of the (a) optical and (b) photoacoustic spectra of NO_2 at 10 torr. The photoacoustic spectrum was obtained using a double-beam spectrometer having carbon black in the reference channel. The dissociation limit is shown by the arrow. (From Harshbarger and Robin, 1973a.)

to the neighborhood of 4000 Å, at which point the absorption peaks and then begins to decrease slowly, whereas the heat amplitude suffers a dramatic reduction to about two-thirds of its peak intensity in the same wavelength region. It is at 4000 Å that the dissociation limit of NO_2 is reached:

$$NO_2 \rightarrow NO(^2\Pi) + O(^3P) \tag{9}$$

Under our high pressure conditions (10 torr) there is no fluorescence. The recombination of NO and O is rather slow, but the O atom can go on to react exothermically with another NO_2 molecule:

$$O + NO_2 \rightarrow NO_2 + O_2 + 47\ \text{kcal mole}^{-1} \tag{10}$$

Energetically then, at the dissociation limit one is investing 72 kcal mole^{-1} in order to produce cold O and NO fragments, which can then react further to release 47 kcal mole^{-1}.* This leads one to expect the amount of heat released at 4000 Å to be diminished by 47/72 of its amplitude prior to dissociation. This would seem to fit the data, Fig. 5, without having to invoke other underlying continua.

Though the dissociation limit of NO_2 is 3986 Å, in fact the fall-off in the photoacoustic spectrum begins at 4200 Å. Mass spectrometric studies also show the dissociation to begin at this wavelength.

The use of a high-resolution dye laser for exciting NO_2 reveals extensive fine structure in the photoacoustic spectrum, but in spite of this, there is still no sharp onset for the dissociation (Kaya *et al.*, 1975b). Angus *et al.* (1975) have used a 250-mW dye laser for determining the presence of NO_2 in air photoacoustically. Using a cell tuned for an acoustic resonance at 1500 Hz, they observed the NO_2 spectrum photoacoustically (5800–6100 Å) in 10 torr of air containing about 10 ppb NO_2. Since the molar extinction coefficient of NO_2 in this region is less than 20, the technique has obvious applications for the analytical determination of not only NO_2, but of any gas having even only weak absorption in the wavelength region of the laser.

B. Iodine

In the case of NO_2 discussed above, there is a significant difference between absorption and photoacoustic spectra due to a thermochemical effect. In iodine vapor, the situation would at first appear to be the same; however, the effect here is kinetic rather than thermochemical. In the visible spectrum of iodine vapor, there are three overlapping excited states: the $^3\Pi_{1u}$ state which is only very slightly bound and dissociates into ground state atoms ($^2P_{3/2}$), the unbound $^1\Pi_{1u}$ state, which also dissociates to $^2P_{3/2}$ atoms, and the $^3\Pi_{0u^+}$ excited state, which is strongly bound and dissociates at ~5000 Å

* A moment's thought will show that the pressure effect resulting from the fission of a molecule into two will be insignificant compared with the heat coming from subsequent relaxation of the fragments.

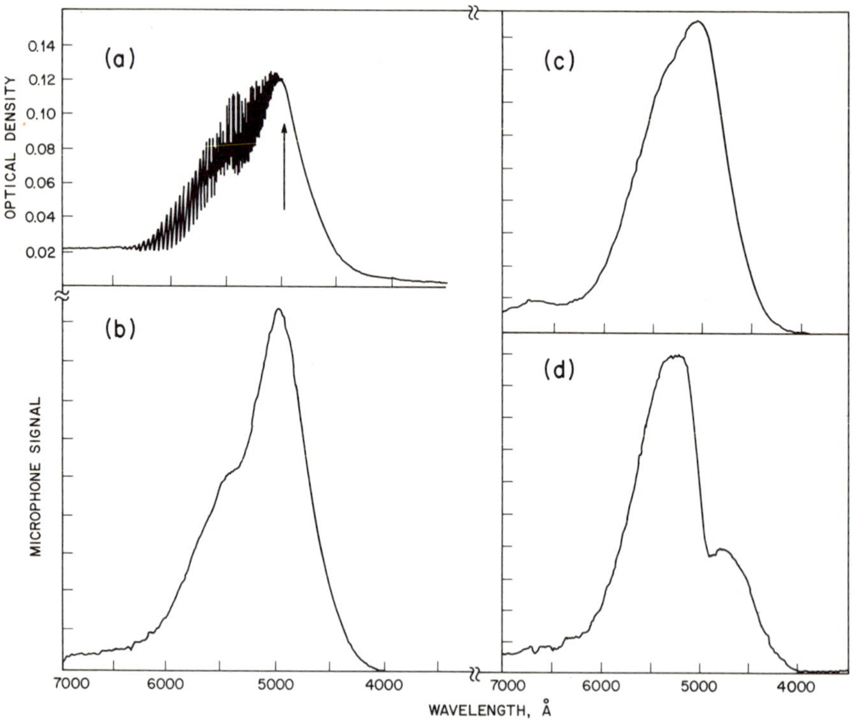

Fig. 6. (a) The optical absorption spectrum of iodine vapor. The dissociation limit is shown by the arrow. (b) The corresponding optoacoustic signal (400 Hz). (c) Optoacoustic signal of I_2 (0.37 torr) and NO (25 torr) gas mixture (400 Hz). (d) Optoacoustic signal of I_2 (0.37 torr) and O_2 (25 torr) gas mixture (400 Hz). (From Harshbarger and Robin, 1973b.)

into one ground state iodine atom and one excited iodine atom ($^2P_{1/2}$). The absorption spectrum to these three states is shown in Fig. 6a with the transition to $^3\Pi_{0u^+}$ being responsible for the long parade of vibronic structure. The photoacoustic spectrum (Fig. 6b) was taken with far less resolution, and so only reflects the envelope of the absorption spectrum (Harshbarger and Robin, 1973b).

When mixed with any laboratory gas, save one, the iodine photoacoustic spectrum remains unchanged. For example, on adding NO (Fig. 6c) essentially the same profile is measured as for the case of pure iodine vapor. However, in the presence of oxygen (Fig. 6d), the photoacoustic profile is altered dramatically; beginning at ~5200 Å, it is clear again that a significant fraction of the absorbed light from that point on is not being converted into heat which can activate the microphone.

The explanation here is the following. Because the wavelength at which the fall-off begins is also the dissociation limit of the $^3\Pi_{0u^+}$ state, it must be that the fall-off is in some way caused by the production of $^2P_{1/2}$ excited atoms and their subsequent fate in the presence of oxygen. The $^2P_{1/2}$ iodine atom is above its ground state by 7603 cm^{-1}, whereas the oxygen molecule has its first excited state ($^1\Delta_g$) at 7886 cm^{-1} above its ground state ($^3\Sigma_g^-$). Hence, it is postulated as a first step that the energy transfer $I(^2P_{1/2}) + O_2(^3\Sigma_g^-) \rightarrow I(^2P_{3/2}) + O_2(^1\Delta_g)$ is operative, being endothermic by only 284 cm^{-1}. Note now that the $^1\Delta_g$ state is remarkably metastable, with a radiative lifetime of over 45 min. Moreover, at the experimental pressure of 25 torr, the $^1\Delta_g$ state of O_2 is relaxed by collisions, but ineffectively, having a lifetime of 0.4 sec in this regard. Now according to Eq. (3), the amplitude in the photoacoustic spectrum will decrease by a factor of over 100 upon crossing 5200 Å, for above this wavelength τ is measured to be $< 10^{-6}$ sec, whereas below it, τ is 0.4 sec when measured at 400 Hz. The damping effect of such a long τ below 5200 Å can only be overcome by going to a low ω of about 1 Hz, which is lower than our equipment can operate. Said differently, the heat is released from $^1\Delta_g$ at a rate ($1/0.4\ sec^{-1}$) which is far too slow to be registered in an experiment operated at a characteristic rate of 400 sec^{-1}.

Note that the heat amplitude in the region of the O_2 trapping is decreased, but not by a factor of 100. The explanation here is that in this spectral region there are underlying continua which do not energy transfer to O_2 since they dissociate into $^2P_{3/2}$ iodine atoms. The estimates of Oldman *et al.* (1971) that the absorption in the region just beyond $^3\Pi_{0u^+}$ dissociation is 50–70% to states giving ground state atoms is quite consistent with the amount of heat found in this region photoacoustically.

Two final points remain to be stressed:

1. The absorption spectra of the I_2/O_2 and I_2/NO gas mixtures are indistinguishable; it is only in their relaxation that they differ, as readily revealed by photoacoustic spectroscopy.
2. Only O_2 is effective in removing heat from the $^3\Pi_{0u^+}$ continuum, since only this gas can accept energy from $I(^2P_{1/2})$ and trap it on a time scale long with respect to $1/\omega$.

There is no O_2 effect in the photoacoustic spectrum of Br_2 at the corresponding dissociation limit, since the $^2P_{1/2}$–$^2P_{3/2}$ interval in the excited Br atom is far too low (3685 cm^{-1}) to excite $^1\Delta_g$ of O_2.

C. *Oxygen*

The long lifetime of the $^1\Delta_g$ state of O_2, which is the essence of the O_2 effect on the iodine spectrum, has been studied directly using excitation to $v' = 1$ of $^1\Delta_g$ via an Nd:YAG laser (Parker and Ritke, 1973). Being careful to work at frequencies far from cell resonances, application of Eq. (3) at constant oxygen pressure P gave $P\tau = 0.05$ atm sec. The same figure was obtained using short-pulse excitation and monitoring the exponential rise of the cell pressure. The lifetime in $^1\Delta_g$ ($v' = 1$) is significantly longer than that for $^1\Delta_g$ ($v' = 0$), but in either case, the relaxation mechanism is not yet understood.

In a very different experiment, Cross and Ardila (1970) illuminate O_2 in the Schumann–Runge bands and beyond. The dissociation to atoms drives a weak shock wave of higher pressure followed by a rarifaction moving with the radiation front in the gas. In this experiment, the heat resulting from relaxation back to the O_2 $^3\Sigma_g^-$ ground state is not a factor.

D. *Ketones*

Because the ketones are so interesting photochemically, it is no surprise that several ketones have been investigated photoacoustically. Encompassing four compounds, this is the group most studied to date.

Immediately upon running a fresh sample of acetaldehyde vapor in the photoacoustic spectrometer, one obtains a faithful representation of the $n \rightarrow \pi^*$ absorption band in the 3500–2700-Å region. However, on a second pass, the spectrum will be changed, with a significant amount of heat being "missing" in the 3200–2700-Å region (65% is missing at 2900 Å). This phenomenon has been studied by de Groot and coworkers (de Groot *et al.*, 1972; Emeis *et al.*, 1974), who find first of all that the "missing heat" is really only slow (lifetime ~2 msec), and so does not appear in the in-phase spectrum adjusted to optimize the fast heat. The out-of-phase spectrum does show the distinct band in the 3200–2700-Å region (Harshbarger and Robin, 1973a), and, further, this band is also observed when operating the lock-in amplifier at twice the chopping frequency, indicating a nonlinear process of some sort. After studying the effects of total pressure, incident light intensity and chopping frequency on the amount of slow heat, it was concluded that following $n \rightarrow \pi^*$ excitation, acetaldehyde rapidly intersystem crosses to a highly excited triplet state, which then energy transfers to a highly excited triplet state of biacetyl formed originally by photodecomposition of the acetaldehyde. This triplet state of biacetyl is unstable and rapidly decays to two acetyl radicals. Up to this point, all of the processes have high quantum yields and occur rapidly. Now however, the acetyl radicals recombine to release heat and form biacetyl, but only slowly owing to their low concentrations at low light intensities. Using this scheme, the Dutch workers derive a recombination rate constant of 2×10^{10} dm^3 mol^{-1} sec^{-1} in good agreement with earlier photochemical measurements. Note that the chemically active triplet state of biacetyl is not the phosphorescing triplet state (T_1), and that though the onset of the missing-heat effect coincides with the onset of S_2

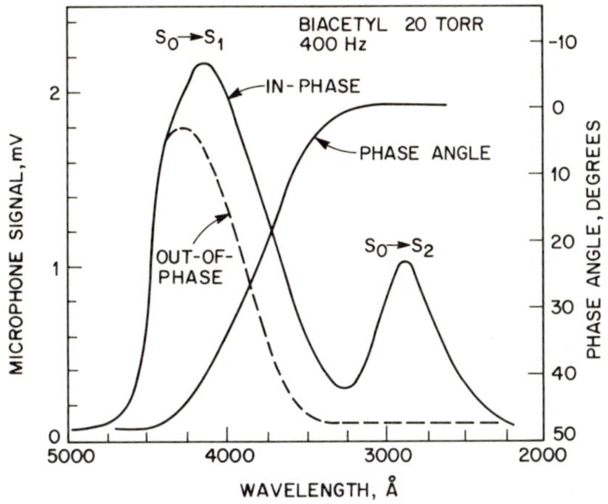

Fig. 7. The in-phase and out-of-phase photoacoustic spectra of biacetyl, taken with a phase angle chosen to extremize the S_2 signal. The phase angle at which the signal is nulled at each wavelength is also shown. (From Kaya *et al.*, 1975a).

absorption in biacetyl, and this state is photochemically active, the heat from this state is fast, not slow (see above).

The biacetyl molecule, being a diketone, displays two $n \rightarrow \pi^*$ transitions in the visible–ultraviolet region, peaked at 4200 ($S_0 \rightarrow S_1$) and 2900 Å ($S_0 \rightarrow S_2$). In Fig. 7, the photoacoustic spectrum of this vapor is reported (Kaya *et al.*, 1974, 1975a); the phase angle was chosen so as to maximize the amplitude at the S_2 peak (the in-phase spectrum), and both bands are clearly seen. Rotating the phase angle by 90° then has the effect of nulling the spectrum at the S_2 maximum, and at every other point in the spectrum which is relaxing at the same rate. This out-of-phase spectrum (Fig. 7) is null across the entire S_2 band profile as well as through the short-wavelength end of the S_1 band, but clearly the long-wavelength end of S_1 is relaxing at a different rate. This is seen more neatly by simply plotting the phase angle, which nulls the signal reading at each point in the spectrum. The phase is constant in the (0, 0) region of S_1 with a value close to 45°, but beyond 4400 Å, it rapidly rises to achieve a constant value across the S_2 band, which is 45° smaller (Fig. 7).

This rapid and continuous change of the phase angle across the $S_0 \rightarrow S_1$ band profile is quite unusual, for in the usual case, the phase angle is constant or slowly varying across a vibronic band. By more conventional techniques it is known that excitation into S_1 is followed by a very rapid intersystem crossing to the triplet manifold with almost unit quantum yield. In the region of the $S_1(0, 0)$, only T_1 is energetically accessible. This state has ${}^3A_u$ symmetry, and as such, has no spin–orbit matrix element that can assist in relaxing it to the ${}^1A_g$ ground state. Hence, the radiationless relaxation from T_1 is slow and permits the radiative process though slow itself ($\tau_p = 1.53$ msec) to achieve a significant quantum yield ($\Phi_p = 0.149$). The slow radiationless step from T_1 is reflected in the phase angle of 49° recorded in the lower S_1 region; according to Eq. (2), this phase angle corresponds to $\tau_{nr} = 0.46$ msec. A value of 0.46 msec is not in very good agreement with the optically determined lifetime of 1.53 msec, but it must be remembered that it was determined using the abbreviated Eq. (3), for which it was assumed that $H/L = 0$. If H/L is not zero, then τ_4 will be larger. Referring to Eq. (2), the measured phase angle and radiative lifetime correspond to $H/L = 0.26$, which agrees nicely with Hunter's value of 0.23, determined at 4360 Å (Hunter and Stock, 1974a).

At 4400 Å, the second triplet state of biacetyl becomes involved in the relaxation and so changes the phase angle. According to MO calculations (Drent *et al.*, 1973), its symmetry is ^3B_g, which is spin–orbit coupled to the ground state and so may relax radiationlessly to the ground state much faster than T_1. Our contention is that as the S_1 ladder is scaled, the intersystem crossing to T_2 becomes more favored, with subsequent rapid relaxation to S_0, bypassing T_1. That T_2 does not relax to T_1 is clear from the rapidly decreasing phase angle as the S_1 band is scanned, and this is paralleled by the rapidly decreasing phosphorescence quantum yield from T_1. In S_2, the relaxation is photochemical largely, rather than via luminescence or radiationless transitions.

The high phosphorescence quantum yield near the $S_1(0,0)$ is efficiently quenched by traces of O_2, most likely in a photooxidation step. This relaxation by reaction is so fast that luminescence cannot compete with it. Similarly, one expects then that in the presence of oxygen, the slow heat would be quenched and only fast heat (i.e., $\phi = 0$) would appear. This is demonstrated in Fig. 8, where the phase angles have been adjusted so that in-phase signal can be thought of as fast heat, whereas the out-of-phase spectrum is that of the slow heat. As expected, one sees that the slow heat appearing near the S_1 origin disappears totally in the presence of oxygen. The addition of azoethane has the same effect on the biacetyl photoacoustic spectrum as oxygen.

In the I_2/O_2 discussion above, it was pointed out that for a slow relaxation, a fast ω can reduce the amplitude significantly (Eq. 3), and that at very low ω, fast and slow processes alike will appear to be fast. Indeed, the biacetyl spectrum at 30 Hz is totally in phase, that is, fast, since everything seems fast at 30 Hz.*

Hunter and Stock (1974a) have also studied the photoacoustic effect in biacetyl, working at one wavelength (4360 Å) and studying the amplitude and phase of the acoustic signal as functions of the chopping rate (100–2000 Hz) and incident light intensity. Both the phase angle and the amplitude of the signal were dependent upon the light intensity, as one would expect for

* According to Eq. (2), relaxations characterized by $\tau = 10^{-3}$ sec and $\tau = 10^{-6}$ sec will have phase angles which differ by 68° at 400 Hz, but only by 16° at 30 Hz.

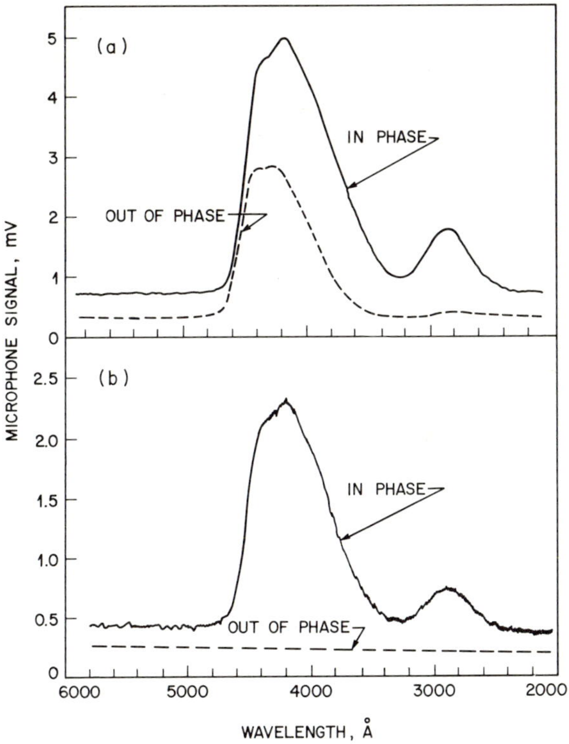

Fig. 8. Comparison of the biacetyl photoacoustic spectra in the presence of nitrogen (a): biacetyl 5.3 torr, nitrogen 4.7 torr, 400 Hz; and oxygen (b): biacetyl 19 torr, oxygen 16 torr, 400 Hz. (From Kaya *et al.*, 1974.)

relaxation resulting from collisions between excited molecules. A triplet–triplet annihilation rate of $2.3 \times 10^{11}\ \mathrm{dm^3\ mol^{-1}\ sec^{-1}}$ was found, which is essentially gas kinetic.

In the $S_0 \rightarrow S_1$ ($n \rightarrow \pi^*$) photoacoustic band of glyoxal, the phase angle shows extraordinary behavior, as seen in Fig. 9 (Kaya *et al.*, 1975b). As the sharp hills and valleys of the absorption spectrum are scanned, the phase angle undergoes corresponding rises and falls, in strong contrast to the monotonic behavior usually observed across a photoacoustic band. The heat amplitude spectrum, when taken with dye-laser excitation, reveals considerable *K*-rotational structure, and this too is clearly reflected in the phase-angle curve. The extremes in phase angle in this band of glyoxal ($\sim 15°$ at 400 Hz) illustrate that the slower parts of the band are relaxing on the tenths-of-millisecond time scale, at the fastest.

The parallelism between the heat amplitude and phase angle in this band of glyoxal is easily shown to be the result of the fact that in this substance, the phase angle at a particular wavelength is dependent upon the energy absorbed

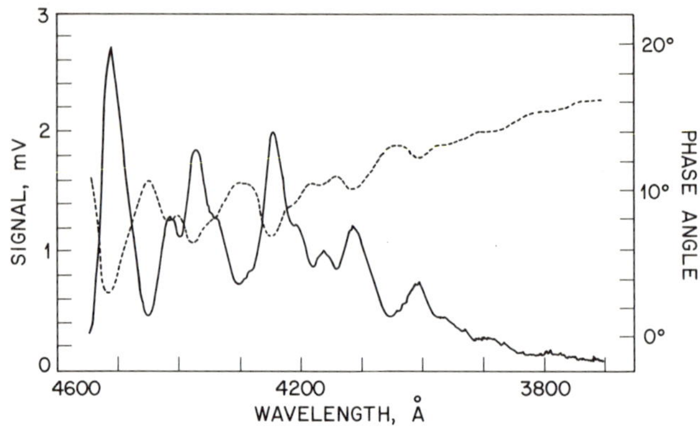

Fig. 9. The amplitude (solid line) and phase angle (dashed line) curves for the $S_0 \rightarrow S_1$ transition of glyoxal at 10 torr and 400 Hz. (From Kaya *et al.*, 1975b.)

at that wavelength. Thus, for example, decreasing the light intensity at the (0, 0) wavelength by a factor of 10 shifts the measured phase angle by 15°. As in the triplet–triplet annihilation work in biacetyl, the intensity-dependent phase angle implies a heat releasing reaction among excited species; indeed, an incident-power/heat-amplitude study at the (0, 0) band shows that this reaction is bimolecular in excited species.

The long lifetimes and slow heats observed following $n \rightarrow \pi^*$ excitation in ketones often result from the S_1 state undergoing intersystem crossing into the triplet manifold, where the molecule is metastable. Indeed, in glyoxal at ~20 torr, the quantum yield for crossing to T_1 is 0.7 (Parmenter, 1964), while at a pressure of 0.2 torr, it is reported to be near unity (Anderson *et al.*, 1973). At the lowest pressures, the radiative lifetime of the T_1 state is about 1 msec. However, this T_1 lifetime is pressure dependent, and at the pressures at which our experiments were run (10–50 torr), the T_1 lifetime is shortened to 2–10 μsec. This decreased T_1 lifetime is a result of the reaction of the T_1 glyoxal with a second ground-state glyoxal molecule to form a well-defined intermediate species of unknown structure, which leads eventually to photochemical products (CO, a polymer, H_2, and possibly H_2CO) (Parmenter, 1964; Yardley, 1972a, b). Clearly, the heat observed photoacoustically on the millisecond time scale is not to be associated with the T_1 self-quenching observed optically on the microsecond time scale, but instead is associated with a process which follows the quenching step.

Our interpretation here is that depending upon the wavelength of excitation, more or less T_1 is produced, and that in about 1 μsec this reacts with an S_0 molecule to produce a new species, M, which may or may not be electronically excited. The species M are relatively long lived and relax via a bimolecular encounter, producing the photoproducts and heat. In such a relaxation scheme, the rate of heat release at each point in the $S_0 \rightarrow S_1$ band will depend directly upon the square of the power absorbed at that wavelength, that is, will be fastest for the most intense bands and slowest for the least intense bands; however, the heat produced comes from the relaxation of the species M, rather than from some excited state of glyoxal.

The aromatic ketones are the final example of photoacoustic spectra in this class of compounds. Among the aromatic ketones, data is available for benzaldehyde, acetophenone and *p*-trifluoromethylacetophenone (Robin and Kuebler, 1975), but only that of acetophenone is discussed here.

In the ultraviolet region (Fig. 10a) there are three transitions in the acetophenone molecule, centered at 3300 Å (S_1, n → π^*), 2750 Å (S_2, $\pi \rightarrow \pi^*$) and 2330 Å (S_3, $\pi \rightarrow \pi^*$). Nulling the photoacoustic spectrum at the center of the S_2 band does not yield a null spectrum; S_1 and S_3 have nonzero amplitudes of opposite sign, and S_2 has different signs on the two sides of the null point. Clearly, the radiationless relaxation rate is varying across the spectrum. A plot of the phase angle (Fig. 10b) reveals a gyrating behavior, not

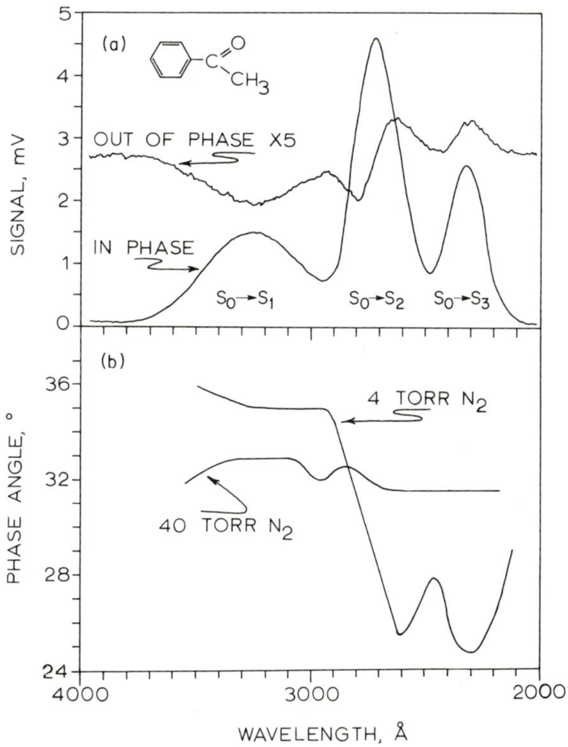

Fig. 10. (a) The in-phase and out-of-phase photoacoustic spectra of acetophenone vapor (0.24 torr) plus 5 torr of helium, at 100 Hz. (b) Phase-angle plots measured for acetophenone vapor (0.24 torr) mixed with 4 torr and 40 torr nitrogen, at 100 Hz. The curve measured with 40 torr nitrogen has been displaced downward by 60°.

all aspects of which are understood. The phase angle is essentially constant in the $n \rightarrow \pi^*$ region, with a value that is 27° larger than the fast S_2 heat in biacetyl. This phase lag at 100 Hz corresponds to a lifetime of 0.8×10^{-3} sec for the slow step of the relaxation chain, and since the quantum yield for heat is 0.5, the nonradiative rate constant is $0.6 \times 10^3\ sec^{-1}$. This is a value agreeing nicely with the $T_1 \rightsquigarrow S_0$ radiationless relaxation rate deduced by Burger and Steel (1975).

At the S_2 threshold, the phase angle drops another 10°, corresponding to a 50% increase in the lifetime of the heat-bearing species, compared to the S_1 region. This dramatic fall off in the heating rate is attributed to the rapid photodecomposition of the S_2 state into C_6H_5CO and CH_3 radicals with high quantum yield, followed by a reaction between radicals, which is slow owing to their low concentrations. As in glyoxal and biacetyl (above), the phase angle in the S_2 region of acetophenone should be intensity dependent, being dependent upon the radical concentration, but this has not been investigated. However S_2 can also be relaxed by collisions that quench the photochemistry. Thus, in the presence of 40 torr of N_2, the slow heat beginning at 2940 Å does not materialize (Fig. 10 lower), but instead, the S_1 heating rate is maintained. This constancy of the phase angle suggests that collisions do not relax S_2 directly to S_0, but rather to S_1 and/or T_1, and thence to S_0. The S_3 region of acetophenone similarly shows a slow heat due to recombination, and this too is quenched by adding a high pressure of N_2.

These S_2 slow heats stemming from radical recombination in acetophenone are not present in the corresponding band of benzaldehyde, where the photochemical decomposition is into stable closed-shell species.

E. Benzene and the Azabenzenes

Long a favorite of absorption spectroscopists, the ${}^1A_g \rightarrow {}^1B_{2u}$ absorption band of benzene beginning at 2600 Å, holds much interest for the photoacoustic spectroscopist as well. This spectrum has been briefly touched upon using photoacoustic spectroscopy in the scanning mode by Kaya *et al.* (1974). In Fig. 11, the heat-amplitude profile of pure benzene vapor is shown; the lack of an out-of-phase signal (at 400 Hz) indicates that all parts of the band are either relaxing at the same rate or at rates faster than $10^6\ sec^{-1}$. However, upon the addition of a trace of biacetyl, a slow-heat signal develops, this being the result of $S_1 \rightsquigarrow T_1$ intersystem crossing in the benzene, followed by energy transfer to the long-lived T_1 state of biacetyl. What is interesting here is that the energy transfer does not extend throughout the ${}^1B_{2u}$ absorption band, but only across the first 3000 cm^{-1}. Thus, the T_1 state of benzene is accessible only from this limited span of S_1, and at points farther into the S_1 spectrum another relaxation process intrudes which is very fast and avoids T_1. The

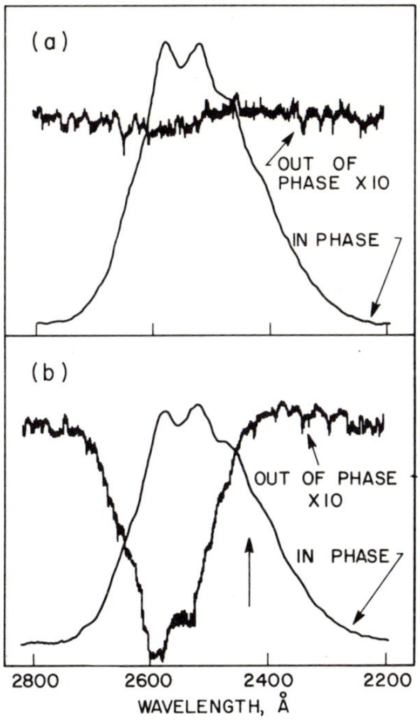

Fig. 11. Photoacoustic spectrum of (a) benzene vapor (68 torr, 400 Hz) in the absence of biacetyl and (b) of benzene vapor (70 torr, 400 Hz) in the presence of 0.3 torr of biacetyl. The arrow in the lower panel indicates the onset of channel 3 relaxation as determined from line-broadening and luminescence studies. (From Kaya *et al.*, 1974).

fast process is the so-called "channel 3" now under active study by many others, and thought to involve interconversion to an isomeric form of benzene. Upon addition of oxygen to the benzene–biacetyl mixture, the slow-heat component is erased, as in direct biacetyl excitation (Fig. 8).

A detailed study on benzene $^1B_{2u}$ relaxation is reported by Hunter and Stock (1973, 1974b). Using excitation outside the area of channel 3 influence (2475 Å), they analyzed the phase angle as a function of pressure and chopping frequency, and their measurements of the rate constants led them to conclude that at high benzene pressures, the T_1 state is relaxed by collision with a ground-state benzene molecule to form a triplet eximer which then decays to two ground-state molecules, whereas at low pressure, excitation at 2475 Å decays by $S_1 \rightsquigarrow S_0$ internal conversion with H/L becoming much larger than 1. As with the iodine spectrum discussed above, the addition of O_2 to S_1-excited benzene results in the trapping of heat in the $^1\Delta_g$ state of the O_2, with the concomitant decrease in heat amplitude. Presumably, this O_2 effect would not be active in the region in which channel 3 functions.

The azabenzene spectra show not only the analog of the benzene transition to $^1B_{2u}$, but also one or more $n \rightarrow \pi^*$ excitations. Corresponding to these, there are additional low-lying triplet levels available for relaxation. The set of azabenzenes from pyridine to *sym*-tetrazine have been studied in the scanning mode by Kaya *et al.* (1975a), with results typified by those obtained with pyrazine.

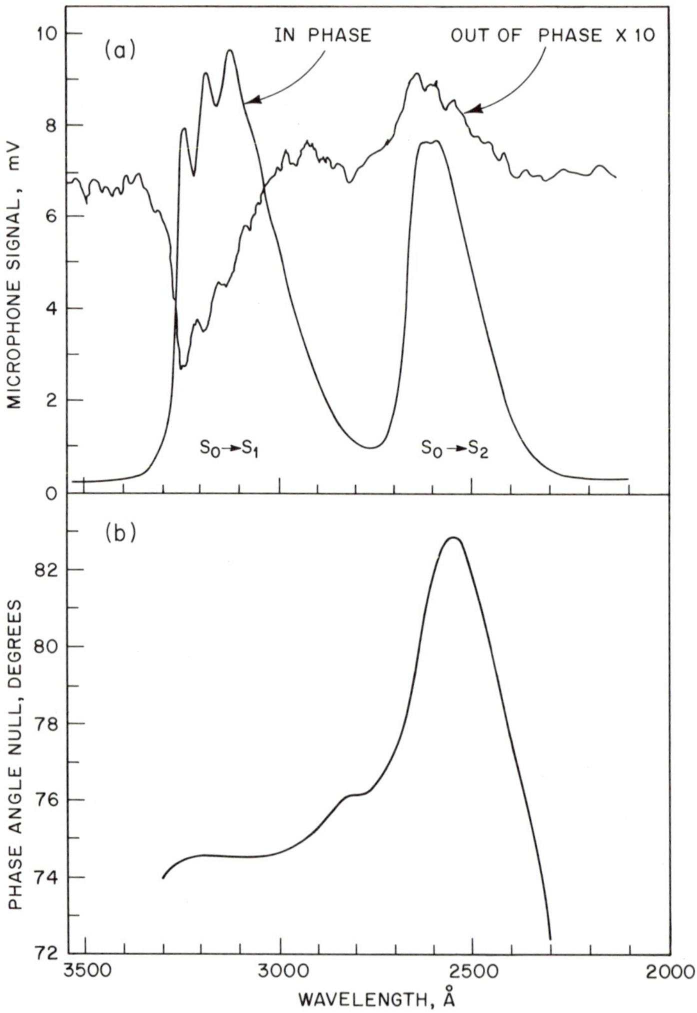

Fig. 12. The in-phase and out-of-phase photoacoustic spectra of pyrazine vapor at 400 Hz (a) and the phase-angle plot (b).

In pyrazine vapor, excitation into the long-wavelength part of S_1 ($n \to \pi^*$) (Fig. 12) yields heat with a 100 msec lifetime, which is the lifetime of the T_1 state. At 2960 Å in S_1, the rate becomes somewhat faster, this being the threshold of $T_3(^3B_{2g})$. The paths from S_1 to S_0 via T_1, T_2, and T_3 are slowed by zero-valued spin–orbit matrix elements, but from S_2 ($\pi \to \pi^*$) to S_0 this is not so, and the relaxation heat is fast. At 2550 Å, the fast heat from S_2 abruptly slows again at pressures near 10 torr, but not at 1 torr. In the presence of O_2, the S_1 and S_2 regions of pyrazine have phase angles differing by less than 1°, whereas in the presence of biacetyl, the S_1 heat becomes very slow and the phase-angle difference is then 16°.

References

Anderson, L. G., Parmenter, C. S., and Poland, H. M. (1973). *Chem. Phys.* **1**, 401.
Angus, A. M., Marinero, E. E., and Colles, M. J. (1975). *Opt. Commun.* **14**, 223.
Bell, A. G. (1881). *Philos. Mag.* **11**, 510.
Burger, M., and Steel, C. (1975). *J. Am. Chem. Soc.* **97**, 4817.
Butcher, R. J., Dennis, R. B., and Smith, S. D. (1975). *Proc. R. Soc. London* **A334**, 541.
Cottrell, T. L., and McCoubrey, J. C. (1961). "Molecular Energy Transfer in Gases." Butterworths, London.
Cottrell, T. L., Macfarlane, I. M., Read, A. W., and Young, A. H. (1966). *Trans. Faraday Soc.* **62**, 2655.
Cross, R. C., and Ardila, R. (1970). *Can. J. Phys.* **48**, 2640.
deGroot, M. S., Emeis, C. A., Hesselman, I. A. M., Drent, E., and Farenhorst, E. (1972). *Chem. Phys. Lett.* **17**, 332.
Delaney, M. E. (1959). *Sci. Progr.* **47**, 459.
Drent, E., van der Werf, R. P., and Kommandeur, J. (1973). *J. Chem. Phys.* **59**, 2061.
Emeis, C. A., Drent, E., Farenhorst, E., Hesselman, I. A. M., and deGroot, M. S. (1974). *Chem. Phys. Lett.* **27**, 17.
Harshbarger, W. R., and Robin, M. B. (1973a). *Acc. Chem. Res.* **6**, 329.
Harshbarger, W. R., and Robin, M. B. (1973b). *Chem. Phys. Lett.* **21**, 462.
Hunter, T. F., and Stock, M. G. (1973). *Chem. Phys. Lett.* **22**, 368.
Hunter, T. F., and Stock, M. G. (1974a). *J. Chem. Soc. Faraday Trans. II* **70**, 1022.
Hunter, T. F., and Stock, M. G. (1974b). *J. Chem. Soc. Faraday Trans. II* **70**, 1028.
Hunter, T. F., Rumbles, D., and Stock, M. G. (1974). *J. Chem. Soc. Faraday Trans. II* **70**, 1010.
Kaya, K., Harshbarger, W. R., and Robin, M. B. (1974). *J. Chem. Phys.* **60**, 4231.
Kaya, K., Chatelain, C. L., Robin, M. B., and Kuebler, N. A. (1975a). *J. Am. Chem. Soc.* **97**, 2153.
Kaya, K., Robin, M. B., and Kuebler, N. A. (1975b). Unpublished results.
Kreuzer, L. B. (1971). *J. Appl. Phys.* **42**, 2934.
Kreuzer, L. B., and Patel, C. K. N. (1971). *Science* **173**, 45.
Max, E., and Rosengren, L. -G. (1974). *Opt. Commun.* **11**, 422.
Oldman, R. J., Sander, R. K., and Wilson, K. R. (1971). *J. Chem. Phys.* **54**, 4127.
Parker, J. G. (1973). *Appl. Opt.* **12**, 2974.
Parker, J. G., and Ritke, D. N. (1973). *J. Chem. Phys.* **59**, 3713.
Parmenter, C. S. (1964). *J. Chem. Phys.* **41**, 658.
Read, A. W. (1967–1968). *Adv. Mol. Relaxation Processes* **1**, 257.
Robin, M. B., and Kuebler, N. A. (1975). *J. Am. Chem. Soc.* **97**, 4822.
Rosengren, L. -G. (1973). *Infrared Phys.* **13**, 109.
Rosengren, L. -G. (1975). *Appl. Opt.* **14**, 1960.
Sessler, G. M., and West, J. E. (1964). *J. Audio Eng. Soc.* **12**, 129.
Sessler, G. M., and West, J. E. (1966). *J. Acoust. Soc. Am.* **40**, 1433.
Stella, G., Gelfand, J., and Smith, W. H. (1975). *Chem. Phys. Lett.* **39**, 146.
Whinnery, J. R. (1974). *Acc. Chem. Res.* **7**, 225.
Yardley, J. T. (1972a). *J. Chem. Phys.* **56**, 6192.
Yardley, J. T. (1972b). *J. Am. Chem. Soc.* **94**, 7283.

8

Solid State Photoacoustic Spectroscopy

Allan Rosencwaig*

Bell Laboratories
Murray Hill, New Jersey

* Present address: Lawrence Livermore Laboratory, Livermore, California.

I. Introduction

It is commonly thought that the photoacoustic effect in gases was discovered before the analogous effect in solids. In actual fact, just the reverse is true. Alexander Graham Bell (1880) gave a report to the American Association for the Advancement of Science on his work on the photophone. In this report, he briefly reported the accidental discovery of what we now term the photoacoustic or optoacoustic effect in solids.

In brief, Bell's photophone consisted of a voice-activated mirror, a selenium cell, and an electrical telephone receiver. A beam of sunlight was intensity modulated at a particular point by means of the voice-activated mirror. This intensity-modulated beam was focused onto a selenium cell, the electrical resistance of which varied with the intensity of light falling on it. The selenium cell was incorporated in a conventional telephone circuit, and thus a voice-modulated beam of light resulted in electrically reproduced telephonic speech.

While experimenting with the photophone, Bell discovered that it was possible to obtain, at times, an audible signal directly, that is, in a nonelectrical fashion. This phenomenon occurred if the beam of sunlight was rapidly interrupted, as with a rotating slotted disk, and then focused on solid substances that were in the form of thin diaphragms connected to a hearing tube.

In another publication, Bell (1881) described in detail his further work on the photoacoustic effect. He found that the need for diaphragms was unnecessary. In particular, he found that if solid matter were placed inside a closed glass tube to which was attached a hearing tube, a quite audible signal could be detected if the material in the tube were then illuminated with a rapidly interrupted beam of sunlight. Bell (1881, p. 515) noted that "the loudest signals are produced from substances in a loose, porous, spongy condition, and from those that have the darkest or most absorbent colours."

In a series of definitive experiments, Bell (1881) demonstrated that the photoacoustic effect in solids was dependent upon the absorption of light, and that the strength of the acoustic signal was in turn dependant on how strongly the incident light was absorbed by the material in the cell. He concluded that "the nature of the rays that produce sonorous effects in different substances depends upon the nature of the substances that are exposed to the beam, and that the sounds are in every case due to those rays of the spectrum that are absorbed by the body" (Bell, 1881, p. 527).

In addition to studying the photoacoustic effect with solids, Bell (1881) and his associate, Sumner Tainter, also studied the effect in liquids and gases. They observed that only weak signals were produced when the cell was filled with a light-absorbing liquid, but that quite strong signals were obtained when the cell was filled with light absorbing gases. Photoacoustic experiments on gases were also performed by John Tyndall (1881) and Wilhelm Röntgen (1881), who had heard about Bell's discovery of the previous year. They found, as did Bell, that the photoacoustic effect in light-absorbing gases could be readily observed for many gases.

After this initial flurry of interest and several subsequent papers by other authors, experimentation with the photoacoustic effect apparently ceased, and was not revived until almost 50 years later, when Viengerov (1938) began using the photoacoustic effect in gases to study light absorption in gases and to evaluate concentrations of gaseous species in gas mixtures. As microphones and their associated electronics became better developed, commercial gas analysis systems, called spectrophones, working on the photoacoustic principle became available. Since the subsequent history of the photoacoustic effect with gases has been extensively covered in earlier chapters of this volume, I shall not pursue this matter here, except to note that it was during this period that the effect, which in 1880 and 1881, had no generally accepted name, now became known as the *optoacoustic effect*. The change of name from "optoacoustic" to "photoacoustic" was instituted by this author to reduce present confusion with the acoustooptic effect in which light interacts with acoustic or elastic waves in a crystal.

At any rate, it is most intriguing, that in spite of the strong rebirth of the photoacoustic effect in 1938, this rebirth was apparently limited to gases. It was not until a few years ago, some 90 years since Bell's original discovery, that the photoacoustic effect in solids and the potential applications of photoacoustic spectroscopy of solids were "rediscovered." How, in spite of the many years of work with gases since 1938, solid-state photoacoustic spectroscopy remained forgotten is a puzzling mystery.

Since 1973, however, the photoacoustic effect with solids has strongly reemerged, with the development of a modern spectroscopic technique that appears to be extremely useful for the investigation of solid and semisolid materials (Maugh, 1975).

Although more is said about experimental methodology later in this chapter, a brief description of present-day techniques might be appropriate. In photoacoustic spectroscopy of solids, or PAS, the sample to be studied is placed inside a closed cell containing a gas, such as air, and a sensitive microphone. The sample then is illuminated with chopped monochromatic light. The analog signal from the microphone is processed by a tuned amplifier the output of which is recorded as a function of the wavelength of the incident light. In this way, photoacoustic spectra are obtained, and these spectra are found to correspond to the true optical absorption spectra of the samples.

One of the principal advantages of photoacoustic spectroscopy (PAS) is that it enables one to obtain spectra, similar to optical absorption spectra, on any type of solid or semisolid material, whether it be crystalline, powder, amorphous, smear, gel, etc. This capability is based on the fact that only the absorbed light is converted to sound. Scattered light, which presents such a serious problem when dealing with many solid materials by conventional spectroscopic techniques, presents no serious difficulties in PAS. Furthermore, it has been found experimentally that good optical absorption data can be obtained, with the photoacoustic technique, on materials that are completely opaque to transmitted light. Photoacoustic spectroscopy has already found important applications in research and analysis of inorganic, organic, and biological solids and semisolids (Rosencwaig, 1973a, b, 1975a, b; Rosencwaig and Hall, 1974, 1975). It furthermore appears to have a strong potential as a tool in surface studies and deexcitation studies (Rosencwaig, 1975a, b). These several applications and potential uses are explored in Sections IV–VI.

II. Theory of the Photoacoustic Effect in Solids

A. *Historical Theories*

Alexander Graham Bell's experiments on the photoacoustic effect with solids, liquids and gases, reported in 1880 and 1881, generated considerable interest at that time. However, the phenomenon was evidently regarded as a curiosity of no great functional or scientific value and was quickly forgotten. Fifty years later, the photoacoustic effect was reborn for the study of gases, and the theory of this effect in gases has since been thoroughly expounded, as noted in the prior chapters of this book. A satisfactory explanation for the photoacoustic effect in solids has been developed only recently, although several attempts to do so have been made both in the last century as well as in this one.

In attempting to account for the audible signal obtained from his experiments with dark spongy solids, such as lampblack, Bell (1881) hypothesized that:

> When a beam of sunlight falls upon the mass, the particles of lampblack are heated, and consequently expand, causing a contraction of the air-spaces or pores among them. Under these circumstances a pulse of air should be expelled, just as we would squeeze out water from a sponge. The force with which the air is expelled must be greatly increased by the expansion of the air itself, due to contact with the heated particles of lampblack. When the light is cut off, the converse process takes place. The lampblack particles cool and contract, thus enlarging the air-spaces among them, and the enclosed air also becomes cool. Under these circumstances a partial vacuum should be formed among the particles, and the outside air would then be absorbed, as water is by a sponge when the pressure of the hand is removed. I imagine that in some such manner as this a wave

of condensation is started in the atmosphere each time a beam of sunlight falls upon the lampblack, and a wave of rarefraction is originated when the light is cut off. (p. 515)

In the case when the illuminated solid is in the form of a thin flexible membrane or disk, Bell supported the theory of Lord Rayleigh (1881), who concluded that the primary source for the photoacoustic signal was the mechanical vibration of the disk resulting from uneven heating of the disk when struck by the beam of sunlight.

Mercadier (1881), who also experimented with the effect at this time, suggested that the sound is due to "vibrating movement determined by the alternate heating and cooling produced by the intermittent radiations, principally in the gaseous layer adhering to the solid surface hit by these radia tions" (p. 410).

In my opinion, Preece (1881) suggested a mechanism which comes closest to the modern theory. He wrote that the photoacoustic effect "is purely an effect of radiant heat, and it is essentially one due to the changes of volume in vapors or gases produced by the degradation and absorption of this heat in a confined space" (p. 517). He further noted that, "Cigars, chips of wood, smoke, or any absorbent surfaces placed inside a closed transparent vessel will, by first absorbing and then radiating heat rays to the confined gas, emit sonorous vibrations" (p. 518).

We have found in our laboratory, from experiments in which we first thoroughly evacuated the photoacoustic cell and then refilled it with non-adsorbing noble gases, and from experiments with two-dimensional solids and other materials with weak surface adsorption properties, that adsorbed gases do not play a significant role in the production of the acoustic signal. Furthermore, it can be readily shown that thermal expansion and contraction of the solid, and in fact any thermally induced mechanical vibrations of the solid are generally too small in magnitude to account for the observed acoustic signal. From both experimental and theoretical considerations, we have concluded that the primary source of the acoustic signal in the photoacoustic cell arises from the periodic heat flow from the solid to the surrounding gas as the solid is cyclically heated by the chopped light. We are thus in agreement with the conjectures of Mercadier and Preece some 90 years ago. Neither Mercadier nor Preece attempted to mathematically develop their hypotheses into an actual theory. It appears that the first attempt at this was by Parker (1973). Parker was performing experiments with the photoacoustic effect in gases, and in the course of these experiments, noticed that a small but measurable signal was being produced in his gas cell, even when the gas within the cell was completely transparent to the incoming light. He determined that this signal resulted from the absorption of light within the quartz windows of his cell, and analyzed his results in terms of a periodic heat flow from the windows to the enclosed gas. In his analysis, however, Parker had to make the rather unrealistic assumption that an anomalously large fraction of his incident light was being absorbed within a very small surface layer of his quartz windows in order to explain his results. The reader is referred to his paper (Parker, 1973) for further details.

B. *Present Theory*

In the following pages we will present a theory (Rosencwaig and Gersho, 1975, 1976) for the photoacoustic effect in solids that assumes, as Mercadier, Preece, and Parker did, that the primary source of acoustic signal results from a periodic heat flow from the solid to the surrounding gas, as the solid is cyclically heated by the absorption of the chopped light. This theory is able to account, in a natural direct fashion, for all of the observed effects in photoacoustic spectroscopy of solids.

1. *Heat Flow Equations*

It is known that any light absorbed by a solid is converted, in part, or in whole into heat by nonradiative deexcitation processes within the solid. We formulate here a one-dimensional model of the heat flow in the cell resulting from this absorbed light energy. Consider a simple cylindrical cell as shown in Fig. 1. The cell has a diameter D and length L. We assume that the length L is small compared to the wavelength of the acoustic signal and the microphone (not shown) will detect the average pressure produced in the cell. The sample is considered to be in the form of a disk having diameter D and thickness l. The sample is mounted so that its front surface is exposed to the gas (air) within the cell and its back surface is against a poor thermal conductor of thickness l_b. The length l_g of the gas column in the cell is then given by $l_g = L - l - l_b$. We further assume that the gas and backing materials are not light absorbing.

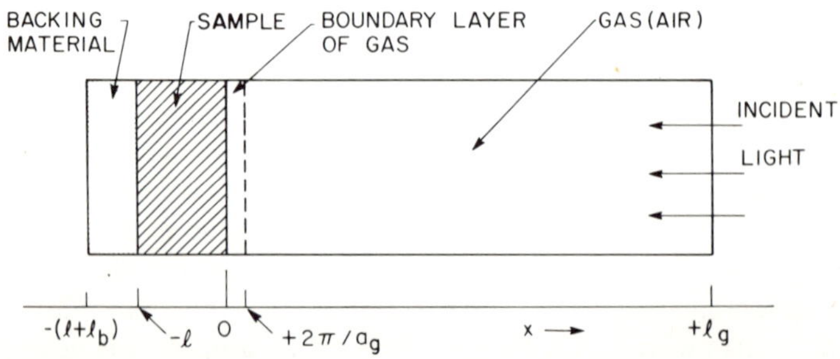

Fig. 1. Cross-sectional view of a simple cylindrical photoacoustic cell.

We define the following parameters: k_j, the thermal conductivity of material j [cal (cm-sec-°C)$^{-1}$]; ρ_j, the density of material j (gm cm^{-3}); C_j, the specific heat of material j [cal (gm-°C)$^{-1}$]; $\beta_j = k_j/\rho_j C_j$, the thermal diffusivity of material j (cm^2 sec^{-1}); $a_j = \sqrt{\omega/2\beta_j}$, the thermal diffusion coefficient of material j (cm^{-1}); and $\mu_j = a_j^{-1}$, the thermal diffusion length of material j (cm), where j can take the subscripts s, g, and b for the solid, gas, and backing material respectively, and ω denotes the chopping frequency of the incident light beam in radians per second.

We assume a sinusoidally chopped monochromatic light source with wavelength λ incident on the solid with intensity

$$I = \tfrac{1}{2} I_0 (1 + \cos \omega t),$$

where I_0 is the incident monochromatic light flux (W cm^{-2}). Let α denote the optical absorption coefficient of the solid sample (in reciprocal centimeters) for the wavelength λ. The heat density produced at any point x owing to light absorbed at this point in the solid is then given by

$$\tfrac{1}{2} \alpha I_0 e^{\alpha x} (1 + \cos \omega t),$$

where x takes on negative values since the solid extends from $x = 0$ to $x = -l$ with the light incident at $x = 0$. Note also from Fig. 1 that the air column extends from $x = 0$ to $x = l_g$ and the backing from $x = -l$ to $x = -(l + l_b)$.

The thermal diffusion equation in the solid taking into account the distributed heat source can be written as

$$\partial^2 \phi/\partial x^2 = \beta_s^{-1}\, \partial \phi/\partial t - A e^{\alpha x}(1 + e^{i\omega t}) \qquad \text{for} \quad -l \le x \le 0, \qquad (1)$$

with $A = \alpha I_0 \eta / 2k_s$, where ϕ is the temperature, and η is the efficiency at which the absorbed light at wavelength λ is converted to heat by the nonradiative deexcitation processes. In this chapter it is assumed that $\eta = 1$, a reasonable assumption for most solids at room temperature. For the backing and the gas, the heat diffusion equations are given by

$$\partial^2 \phi/\partial x^2 = \beta_b^{-1}\, \partial \phi/\partial t, \qquad -l - l_b \le x \le -l \qquad (2)$$

$$\partial^2 \phi/\partial x^2 = \beta_g^{-1}\, \partial \phi/\partial t, \qquad 0 \le x \le l_g. \qquad (3)$$

A more exact treatment for the gas can be derived using equations of fluid dynamics. However, the approximate treatment given here gives results consistent with the more exact method, and has the advantage of presenting a simpler physical description of the process. The real part of the complex valued solution $\phi(x, t)$ of Eqs. (1)–(3) is the solution of physical interest and represents the temperature in the cell relative to ambient temperature as a function of position and time. Thus, the actual temperature field in the cell is given by

$$T(x, t) = \mathrm{Re}\{\phi(x, t)\} + \Phi,$$

where Re denotes the "real part of" and Φ is the ambient (room) temperature.

To completely specify the solution of Eqs. (1)–(3), the appropriate boundary conditions are obtained from the requirement of temperature and heat flux continuity at the boundaries $x = 0$ and $x = -l$, and from the constraint that the temperature at the cell walls $x = +l_g$ and $x = -l - l_b$ is at ambient temperature. The latter constraint is a reasonable assumption for metallic cell walls but in any case it does not affect the ultimate solution for the acoustic pressure.

Finally, we note that we have assumed the dimensions of the cell are small enough to ignore convective heat flow in the gas at steady-state conditions.

2. *Temperature Distribution in the Cell*

The general solution for $\phi(x, t)$ in the cell neglecting transients can be written as

$$\phi(x,t) = \begin{cases} l_b^{-1}(x + l + l_b)W_0 + W\exp[\sigma_b(x + l) + i\omega t], & -l - l_b \leq x \leq -l \\ e_1 + e_2 x + de^{\alpha x} + (Ue^{\sigma_s x} + Ve^{-\sigma_s x} - Ee^{\alpha x})e^{i\omega t}, & -l \leq x \leq 0 \\ (1 - x/l_g)\theta_0 + \theta\exp(-\sigma_g x + i\omega t), & 0 \leq x \leq l_g, \end{cases} \tag{4}$$

where W, U, V, E, and θ are complex-valued constants, e_1, e_2, d, W_0, and θ_0 are real-valued constants, and $\sigma_j = (1 + i)a_j$ with $a_j = \sqrt{\omega/2\beta_j}$. In particular it should be noted that θ and W represent the complex amplitudes of the periodic temperatures at the sample-gas boundary ($x = 0$) and the sample-backing boundary ($x = -l$) respectively. The dc solution in the backing and gas already make use of the assumption that the temperature (relative to ambient) is zero at the ends of the cell. The quantities W_0 and θ_0 denote the dc component of the temperature (relative to ambient) at the sample surfaces $x = -l$ and $x = 0$, respectively. The quantities E and d, determined by the forcing function in Eq. (1), are given by

$$d = -A/\alpha^2 \tag{5a}$$

$$E = A/(\alpha^2 - \sigma_s^2) = \alpha I_0/2k_s(\alpha^2 - \sigma_s^2) \tag{5b}$$

In the general solution (4) we have omitted the growing exponential component of the solutions to the gas and backing material, because for all frequencies ω of interest the thermal diffusion length is small compared with the length of the material in both the gas and the backing. That is, $\mu_b \ll l_b$ and $\mu_g \ll l_g$ ($\mu_g \sim 0.02$ cm for air when $\omega = 630$ rad sec^{-1}), and hence the sinusoidal components of these solutions are sufficiently damped so that they are effectively zero at the cell walls. Therefore, the growing exponential components of the solutions would have coefficients that are essentially zero in order to satisfy the temperature constraint at the cell walls.

The temperature and flux continuity conditions at the sample surfaces are explicitly given by

$$\phi_g(0,t) = \phi_s(0,t), \tag{6a}$$
$$\phi_b(-l,t) = \phi_s(-l,t), \tag{6b}$$
$$k_g(\partial\phi_g/\partial x)(0,t) = k_s(\partial\phi_s/\partial x)(0,t), \tag{6c}$$
$$k_b(\partial\phi_b/\partial x)(-l,t) = k_s(\partial\phi_s/\partial x)(-l,t), \tag{6d}$$

where the subscripts s, b, and g identify the solution to Eqs. (4) for the temperature in the solid, backing, and gas, respectively. These constraints apply separately to the d.c. component and the sinusoidal component of the solution. From Eqs. (6), we obtain for the dc components of the solution

$$\theta_0 = e_1 + d, \tag{7a}$$
$$W_0 = e_1 - e_2 l + de^{-\alpha l}, \tag{7b}$$
$$-(k_g/l_g)\theta_0 = k_s e_2 + k_s \alpha d, \tag{7c}$$
$$(k_b/l_b)W = k_s e_2 + k_s \alpha d e^{-\alpha l}. \tag{7d}$$

Equations (7) determine the coefficients e_1, e_2, W_0, and θ_0 for the time-independent (dc) component of the solution. Applying Eqs. (6) to the sinusoidal component of the solution, yields

$$\theta = U + V - E, \tag{8a}$$
$$W = e^{-\sigma_s l}U + e^{\sigma_s l}V - e^{-\alpha l}E \tag{8b}$$
$$-k_g\sigma_g\theta = k_s\sigma_s U - k_s\sigma_s V - k_s\alpha E, \tag{8c}$$
$$k_b\sigma_b W = k_s\sigma_s e^{-\sigma_s l}U - k_s\sigma_s e^{\sigma_s l}V - k_s\alpha e^{-\alpha l}E. \tag{8d}$$

These quantities, together with the expression for E in Eq. (5b), determine the coefficients U, V, W, and θ. Hence, the solutions to Eqs. (7) and (8) allow us to evaluate the temperature distribution (Eqs. 4) in the cell in terms of the optical, thermal, and geometric parameters of the system. The explicit solution for θ, the complex amplitude of the periodic temperature at the solid–gas boundary ($x = 0$), is given by

$$\theta = \frac{\alpha I_0}{2k_s(\alpha^2 - \sigma_s^2)}\left\{\frac{(r-1)(b+1)e^{\sigma_s l} - (r+1)(b-1)e^{-\sigma_s l} + 2(b-r)e^{-\alpha l}}{(g+1)(b+1)e^{\sigma_s l} - (g-1)(b-1)e^{-\sigma_s l}}\right\}, \tag{9}$$

where

$$b = k_b a_b / k_s a_s, \tag{10}$$
$$g = k_g a_g / k_s a_s, \tag{11}$$
$$r = (1 - i)\alpha / 2a_s, \tag{12}$$

and as stated earlier, $\sigma_s = (1 + i)a_s$. Thus, Eq. (9) can be evaluated for specific parameter values yielding a complex number whose real and imaginary parts θ_1 and θ_2, respectively, determine the in-phase and quadrature components of the periodic temperature variation at the surface $x = 0$ of the sample. Specifically, the actual temperature at $x = 0$ is given by

$$T(0, t) = \Phi + \theta_0 + \theta_1 \cos \omega t - \theta_2 \sin \omega t,$$

where Φ is the ambient temperature at the cell walls and θ_0 is the increase in temperature owing to the steady-state component of the absorbed heat.

3. *Production of the Acoustic Signal*

As stated in the introduction, it is our contention that the main source of the acoustic signal arises from the periodic heat flow from the solid to the surrounding gas. The periodic diffusion process produces a periodic temperature variation in the gas as given by the sinusoidal (ac) component of the solution (Eqs. 4),

$$\phi_{ac}(x, t) = \theta \exp(-\sigma_g x + i\omega t). \tag{13}$$

Taking the real part of Eq. (13), we see that the actual physical temperature variation in the gas is

$$T_{ac}(x, t) = e^{-a_g x}\{\theta_1 \cos(\omega t - a_g x) - \theta_2 \sin(\omega t - a_g x)\}, \tag{14}$$

where θ_1 and θ_2 are the real and imaginary parts of θ, as given by Eq. (9). As can be seen in Fig. 2, the time-dependent component of the temperature in the gas attenuates rapidly to zero with increasing distance from the surface of the solid. At a distance of only $2\pi/a_g = 2\pi\mu_g$, where μ_g is the thermal diffusion length, the periodic temperature variation in the gas is effectively fully damped out. Thus we can define a boundary layer, as shown in Fig. 1, the thickness of which is $2\pi\mu_g$ (≈ 0.1 cm at $\omega/2\pi = 100$ Hz) and maintain to a good approximation that only this thickness of gas is capable of responding thermally to the periodic temperature at the surface of the sample.

The spatially averaged temperature of the gas within this boundary layer as a function of time can be determined by evaluating

$$\bar{\phi}(t) = (2\pi\mu_g)^{-1} \int_0^{2\pi\mu_g} \phi_{ac}(x, t)\, dx.$$

From Eq. (13) this gives

$$\bar{\phi}(t) \cong (2\sqrt{2\pi})^{-1} \theta \exp[i(\omega t - \pi/4)], \tag{15}$$

using the approximation $e^{-2\pi} \ll 1$.

Because of the periodic heating of the boundary layer, this layer of gas expands and contracts periodically and thus can be thought of as acting as an acoustic piston on the rest of the gas column, producing an acoustic pressure

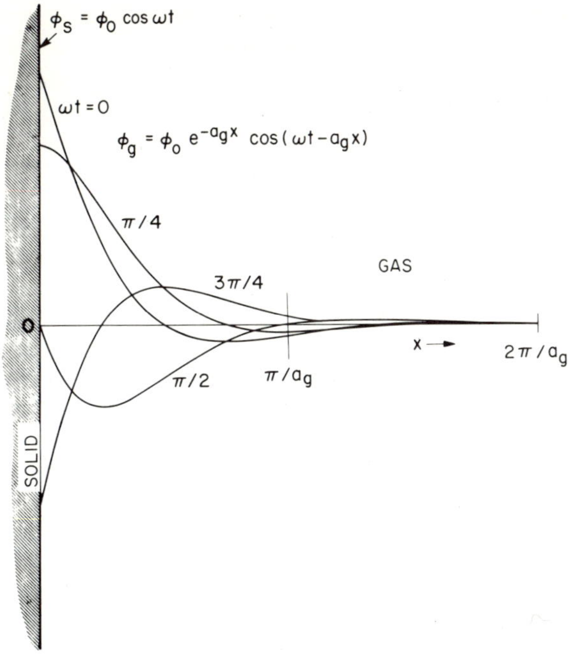

Fig. 2. Spatial distribution of the time-dependent temperature within the gas layer adjacent to the solid surface.

signal that travels through the entire gas column. A similar argument has been used successfully to account for the acoustic signal produced when a conductor in the form of a thin flat sheet is periodically heated by an ac electrical current (Arnold and Crandall, 1917).

The displacement of this gas piston owing to the periodic heating can be simply estimated by using the ideal gas law,

$$\delta_x(t) = 2\pi\mu_g \frac{\bar{\phi}(t)}{T_0} = \frac{\theta\mu_g}{\sqrt{2}\,T_0} \exp[i(\omega t - \pi/4)], \tag{16}$$

where we have set the average dc temperature of this gas boundary layer equal to the dc temperature at the solid surface, $T_0 = \Phi + \theta_0$. Equation (16) is a reasonable approximation to the actual displacement of the layer since since $2\pi\mu_g$ is only ~ 0.1 cm for $\omega/2\pi = 100$ Hz and even smaller for higher frequencies.

If we assume that the rest of the gas responds to the action of this piston adiabatically, then the acoustic pressure in the cell resulting from the displacement of this gas piston is derived from the adiabatic gas law

$$PV^\gamma = \text{constant}$$

where P is the pressure, V the gas volume in the cell, and γ the ratio of the specific heats. Thus, the incremental pressure is

$$\delta P(t) = (\gamma P_0/V_0)\,\delta V = (\gamma P_0/l_g)\,\delta x(t)$$

where P_0 and V_0 are the ambient pressure and volume, respectively, and $-\delta V$ is the incremental volume. Then from Eq. (16),

$$\delta P(t) = Q \exp i(\omega t - \pi/4)], \tag{17}$$

where

$$Q = \gamma P_0 \theta/\sqrt{2} l_g a_g T_0. \tag{18}$$

Thus, the actual physical pressure variation, $\Delta P(t)$, is given by the real part of $\delta P(t)$ as

$$\Delta P(t) = Q_1 \cos(\omega t - \pi/4) - Q_2 \sin(\omega t - \pi/4) \tag{19}$$

or

$$\Delta P(t) = q \cos(\omega t - \psi - \pi/4), \tag{20}$$

where Q_1 and Q_2 are the real and imaginary parts of Q, and q and $-\psi$ are the magnitude and phase of Q, that is,

$$Q = Q_1 + iQ_2 = qe^{-i\psi}.$$

Thus, Q specifies the complex envelope of the sinusoidal pressure variation. Combining Eqs. (9) and (18) we get the explicit formula

$$Q = \frac{\alpha I_0 \gamma P_0}{2\sqrt{2} k_s l_g a_g T_0(\alpha^2 - \sigma_s{}^2)} \times \left\{ \frac{(r-1)(b+1)e^{\sigma_s l} - (r+1)(b-1)e^{-\sigma_s l} + 2(b-r)e^{-\alpha l}}{(g+1)(b+1)e^{\sigma_s l} - (g-1)(b-1)e^{-\sigma_s l}} \right\}, \tag{21}$$

where $b = k_b a_b/k_s a_s$, $g = k_g a_g/k_s a_s$, $r = (1-i)\alpha/2a_s$, and $\sigma_s = (1+i)a_s$ as previously defined. At ordinary temperatures $T_0 \approx \Phi$ so that the dc components of the temperature distribution need not be evaluated. Thus, Eq. (21) may be evaluated for the magnitude and phase of the acoustic pressure wave produced in the cell by the photoacoustic effect.

4. *Special Cases*

The full expression for $\delta P(t)$ is somewhat difficult to interpret because of the complicated expression for Q as given by Eq. (21). However, physical insight may be gained by examining special cases in which the expression for Q becomes relatively simple. We group these cases according to the optical opaqueness of the solids as determined by the relation of the optical absorption length, to the thickness l of the solid. For each category of optical opaqueness, we then consider three cases according to the relative magnitude of the thermal diffusion length μ_s, as compared with the physical length l and the optical absorption length $\mu_\alpha = 1/\alpha$. For all of the cases evaluated below, we make use of the reasonable assumption that $g < b$ and that $b \sim 1$, that is $k_g a_g < k_b a_b$ and $k_b a_b \sim k_s a_s$.

The six cases are illustrated in Fig. 3. It is convenient to define

$$Y = \gamma P_0 I_0 / 2\sqrt{2}\, l_g T_0, \tag{22}$$

which always appears in the expression for Q as a constant factor.

(*a*) *Case 1. Optically Transparent Solids* ($\mu_\alpha > l$) In these cases, the light is absorbed throughout the length of the sample.

Case 1a. Thermally Thin Solids ($\mu_s \gg l$; $\mu_s > \mu_\alpha$) Here we set $e^{-\alpha l} = 1 - \alpha l$; $e^{\pm \sigma_s l} \cong 1$; and $|r| > 1$ in Eq. (21). We then obtain

$$Q = \frac{lY}{2a_g a_b k_b}(\alpha - 2a_s b - i\alpha) \cong \frac{(1-i)\alpha l}{2a_g}\left(\frac{\mu_b}{k_b}\right) Y. \tag{23}$$

The acoustic signal is thus proportional to αl, and since μ_b / a_g is proportional to ω^{-1}, the acoustic signal has a ω^{-1} dependence. For this thermally thin case of $\mu_s \gg l$, the thermal properties of the backing material come into play in the expression for Q.

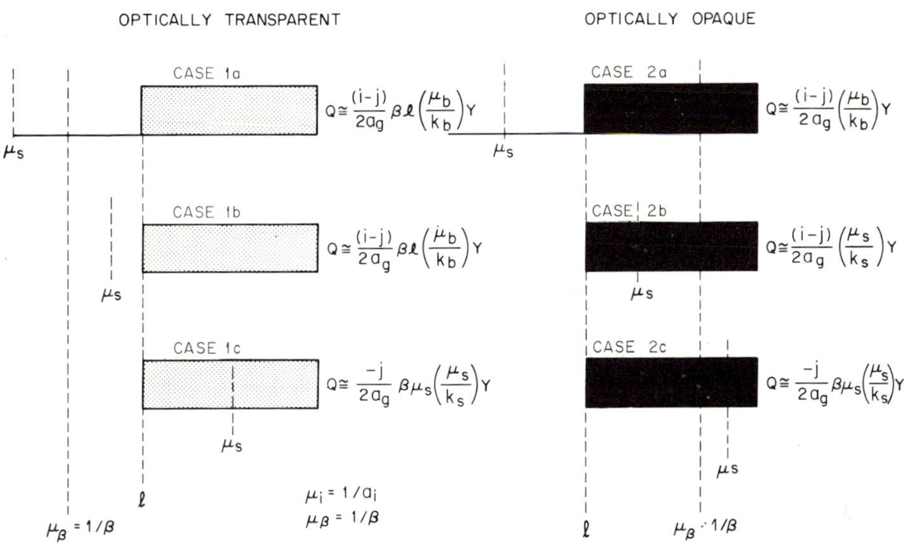

Fig. 3. Schematic representation of special cases for the photoacoustic effect in solids.

Case 1b. Thermally Thin Solids ($\mu_s > l$; $\mu_s < \mu_\alpha$) Here we set $e^{-\alpha l} \cong 1 - \alpha l$, $e^{\pm\sigma_s l} \cong (1 \pm \sigma_s l)$, and $|r| < 1$ in Eq. (21). We then obtain

$$Q = \frac{\alpha l Y}{4k_s a_g a_s{}^3 b}\{(\alpha^2 + 2a_s{}^2) + i(\alpha^2 - 2a_s{}^2)\} \simeq \frac{(1-i)\alpha l}{2a_g}\left(\frac{\mu_b}{k_b}\right)Y. \quad (24)$$

The acoustic signal is again proportional to αl, varies as ω^{-1}, and depends on the thermal properties of the backing material. Equation (24) is identical to Eq. (23).

Case 1c. Thermally Thick Solids ($\mu_s < l$; $\mu_s \ll \mu_\alpha$) In Eq. (21) we set $e^{-\alpha l} \cong 1 - \alpha l$ and $e^{\pm\sigma_s l} \cong 0$ and $|r| \ll 1$. The acoustic signal then becomes

$$Q \simeq -i(\alpha\mu_s/2a_g)(\mu_s/k_s)Y. \quad (25)$$

Here the signal is now proportional to $\alpha\mu_s$ rather than αl. That is, only the light absorbed within the first thermal diffusion length μ_s contributes to the signal, in spite of the fact that light is being absorbed throughout the length l of the solid. Also since $\mu_s < l$, the thermal properties of the backing material present in Eq. (24) are replaced by those of the solid. The frequency dependence of Q in Eq. (25) varies as $\omega^{-3/2}$.

Cases 1a–1c for the so-called "optically transparent" sample demonstrate a unique capability of photoacoustic spectroscopy, to wit, the capability of determining a depth profile of optical absorption within a sample. That is, by starting at a high chopping frequency one will obtain optical absorption information from only a layer of material near the surface of the solid. For materials with low thermal diffusivity this layer can be as small as 0.1 μ at chopping frequencies of 10,000–100,000 Hz. Then by decreasing the chopping frequency, one increases the thermal diffusion length and obtains optical absorption data further within the material, until at ~20 Hz one can obtain data down to 10–100 μm, for materials with low thermal diffusivities and up to 1–10 mm for materials with high thermal diffusivities.

This capability for doing depth profile analysis is quite unique and opens up exciting possibilities in studying layered and amorphous materials, and for determining overlay and thin film thickness.

(b) Case 2. Optically Opaque Solids ($\mu_\alpha \ll l$) In these cases, most of the light is being absorbed within a distance small compared to l.

Case 2a. Thermally Thin Solids ($\mu_s \gg l$; $\mu_s \gg \mu_\alpha$) In Eq. (21) we set $e^{-\alpha l} \cong 0$; $e^{\pm\sigma_s l} \cong 1$; and $|r| \gg 1$. We then obtain

$$Q \cong [(1-i)/2a_g](\mu_b/k_b)Y. \quad (26)$$

In thise case, we have photoacoustic "opaqueness" as well as optical opaqueness, in the sense that our acoustic signal is independent of α. This would be the case of a very black absorber such as carbon black. The signal is quite strong, (it is $(\alpha l)^{-1}$ times as strong as that in Case 1a), and depends on the thermal properties of the backing material, and varies as ω^{-1}.

Case 2b. Thermally Thick Solids ($\mu_s < l$; $\mu_s > \mu_\alpha$) In Eq. (21) we set $e^{-\alpha l} \cong 0$; $e^{\pm \sigma_s l} \cong 0$, and $|r| > 1$. We obtain

$$Q \simeq \frac{Y}{2a_g a_s k_s \alpha}(\alpha - 2a_s - i\alpha) \cong \frac{(1-i)}{2a_g}\left(\frac{\mu_s}{k_s}\right) Y. \tag{27}$$

Equation (27) is analogous to Eq. (26), but the thermal parameters of the backing are now replaced by those of the solid. Again the acoustic signal is independent of α and varies as ω^{-1}.

Case 2c. Thermally Thick Solids ($\mu_s \ll l$; $\mu_s < \mu_\alpha$) We set $e^{-\alpha l} \cong 0$; $e^{\pm \sigma_s l} \cong 0$; and $|r| < 1$ in Eq. (21). We obtain

$$Q = \frac{-i\alpha Y}{4a_g a_s^3 k_s}(2a_s - \alpha + i\alpha) = \frac{-i\alpha\mu_s}{2a_g}\left(\frac{\mu_s}{k_s}\right) Y.$$

This is a very interesting and important case. Optically we are dealing with a very opaque solid (αl, $\gg 1$). However, as long as $\alpha\mu_s < 1$ (i.e., $\mu_s < 1/\alpha$), this solid is not photoacoustically opaque, since, as in Case 1c, only the light absorbed within the first thermal diffusion length μ_s will contribute to the acoustic signal. Thus, even though, this solid is optically opaque, the photoacoustic signal will be proportional to $\alpha\mu_s$. As in Case 1c, the signal is also dependent on the thermal properties of the solid and varies as $\omega^{-3/2}$.

5. *Conclusions*

A theoretical analysis of the photoacoustic effect with solids has been performed. In this analysis, we have assumed that the primary source of the acoustic signal arises from the periodic heat flow from the solid to the surrounding gas. This periodic heat flow causes an oscillatory motion of a thin layer of gas at the solid–gas boundary, and it is this motion of the gas layer that produces the acoustic signal detected in the photoacoustic cell. We have derived the exact solutions for the acoustic pressure produced in the cell due to this process, and have evaluated explicit formulas for certain cases of physical significance. The formulas developed for the special cases have been found to give numerical results for magnitude and phase of the acoustic signal that are in good agreement with the experimental data. This agreement supports our explanation for the mechanism underlying the photoacoustic effect with solids.

Our formulas show that the photoacoustic signal is ultimately governed by the magnitude of the thermal diffusion length of the solid. Thus, even when a solid is optically opaque, it is not necessarily opaque photoacoustically, and, in fact, as long as $\alpha\mu_s < 1$, the photoacoustic signal will be proportional to α, even though the optical thickness, αl, of the sample may be much greater than

1. Since the thermal diffusion length μ_s can be changed by changing the chopping frequency ω, it is therefore possible, with the photoacoustic technique, to obtain optical absorption spectra on all but the most highly opaque solids. This capability of the PAS technique, together with its insensitivity to scattered light, makes its use as a spectroscopic tool for the investigation of solid and semisolid materials highly attractive. In particular, these features give the photoacoustic technique a unique potential for noninvasive in-vivo studies of human tissues, a potential that may have important implications in biological and medical research and in medical diagnostics.

Although it is commonly assumed that the photoacoustic technique will be used to obtain optical absorption data, it should be stressed that other types of measurement can also be performed. It is clear from the above discussion, and from the theory, that as long as two of the three "length" parameters are known, or are varied in a controlled fashion, information about the third parameter can be obtained.

Thus, in the most common class of photoacoustic experiments, one can obtain information about the optical path length $\mu_\alpha = 1/\alpha$ and thereby about the optical absorption coefficient α, as a function of wavelength, by keeping l and μ_s constant. The bulk of the present work in photoacoustic spectroscopy is concerned with this particular class of measurements. These experiments have been quite fruitful, yielding valuable spectroscopic data on inorganic, organic, and biological systems, data which could not be obtained readily by more conventional techniques.

There are, in addition, two other classes of experiments which are of considerable interest and for which the photoacoustic technique is uniquely suited. It is possible to obtain information about the thermal conductivity of a material by measuring the thermal diffusion length μ_s through knowledge of l and μ_α. The thermal conductivity is an important physical parameter which is often very difficult to measure, particularly on powders, amorphous materials, and biological samples. Finally, one can measure l, through a knowledge of μ_α and μ_s, or by keeping μ_α constant and changing μ_s in a controlled fashion, or vice versa. Such an experiment can be extremely useful for measuring the thickness of thin films on substrates that possess different optical or thermal properties than those of the films.

III. Experimental Methodology

As in other forms of spectroscopy, a photoacoustic spectrometer is composed of three main parts: a source of incoming radiation, the experimental chamber, and the data acquisition system. A typical photoacoustic spectrometer is shown in block diagram fashion in Fig. 4.

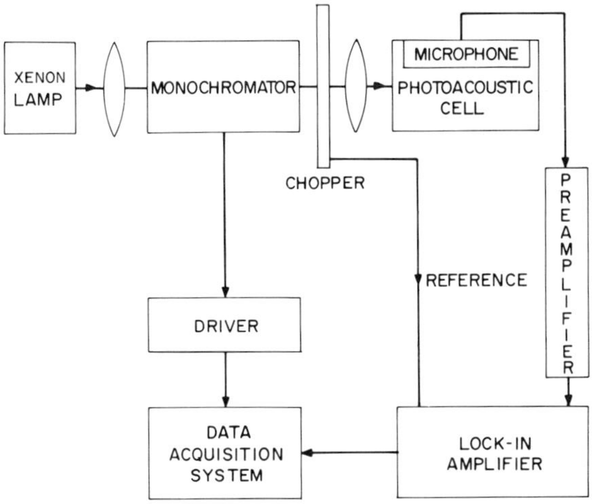

Fig. 4. Block diagram of a typical single-beam photoacoustic spectrometer.

A. Radiation Sources

The most common and still most versatile sources of optical radiation in the ultraviolet, visible, and infrared regions, are provided by the conventional light sources. These are the arc lamp for the ultraviolet–visible, the incandescent lamp for the visible and near infrared, and the glow bar for the mid- to far-infrared regions. Since all three light sources provide strong, broad-band optical radiation, they must be used in conjunction with suitable monochromators. Since the signal-to-noise ratio in photoacoustic spectroscopy increases linearly with the amount of light falling on the sample (Rosencwaig, 1973b), it is clear that one desires an intense light source and a high light throughout (i.e., low f number) monochromator. These light sources generally operate in a continuous mode, and thus a light chopper, usually electromechanical in nature, must be used. This chopper can be located before or after the monochromator. In the theory of the photoacoustic effect in solids (Section II), we have shown that optical absorption spectra on completely opaque samples can be obtained, and that absorption versus depth studies can be performed if one is able to adjust the chopping frequency. Thus, a variable-speed chopper is recommended for optimal versatility.

Another source of optical radiation that can be used in photoacoustic spectroscopy of solids is the laser. A laser requires no monochromator and, if operated in a pulsed mode, would also require no chopper. In the visible wavelength region, dye lasers provide an intense highly monochromatic light readily tunable over a fairly large wavelength range. Dye lasers can also be used with reasonable intensity in the ultraviolet region with the aid of frequency-doubling crystals. Scanning with dye lasers in the ultraviolet region is however tedious at present. In the infrared there are currently no continuously tunable lasers that cover a wide spectral range, although if the experiment can be performed over a narrow wavelength region, then a discrete infrared laser (e.g., the CO or CO_2 lasers) or a tunable spin-flip Raman laser can be used to great advantage to provide intense, highly monochromatic radiation.

B. *Experimental Chamber*

The experimental chamber is the section containing the photoacoustic cell or cells, and all the required optics. The actual design of this chamber will vary depending on whether one is using a single-beam system employing only one photoacoustic cell, or a double-beam system containing two cells, with appropriate beam-splitting optics. The photoacoustic cells will generally incorporate a suitable microphone with its preamplifier. Both a conventional condenser microphone with external biasing, and the electret microphones with internal self-biasing provided from a charged electret foil, are good microphones to use.

Some criteria governing the actual design of the photoacoustic cell are

(a) Acoustic isolation from the outside world.

(b) Minimization of extraneous photoacoustic signal arising from the interaction of the light beam with the walls, windows and the microphone in the cell.

(c) Microphone configuration.

(d) Acoustical means for maximizing the acoustic signal within the cell.

(e) The requirements set by the samples to be studied and the type of experiments to be performed.

Let us now consider the above criteria in more detail:

(a) We have found that the problem of acoustic isolation is not particularly serious providing one uses lock-in detection methods for analyzing the microphone signal. One should, of course use chopping frequencies different from those present in the acoustic and vibrational spectrum of the environment. In addition the cell should be designed with good acoustic seals and with walls of sufficient thickness to form a good acoustic barrier. Some reasonable precautions to isolate the photoacoustic cell from room vibrations should also be taken.

(b) To minimize any photoacoustic signal that may arise from the interaction of the light beam with the walls and windows of the cell, one should employ windows as optically transparent as possible for the wavelength

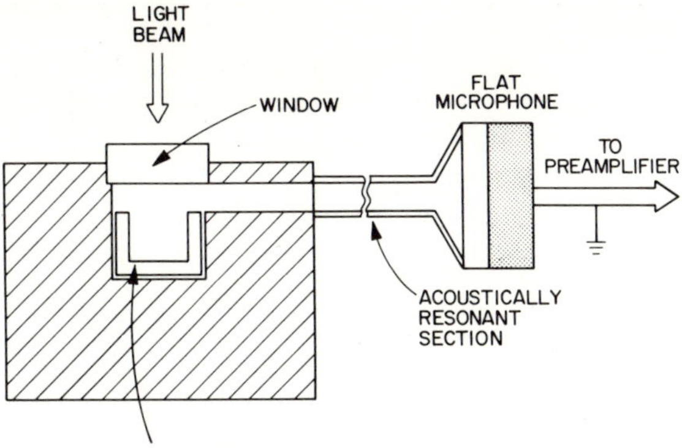

Fig. 7. A photoacoustic cell with an acoustically resonant section.

lower gas temperatures. All these methods increase the photoacoustic signal without limiting the choice of chopping frequencies.

(e) How closely one adheres to the above criteria of course depends on the type of sample used (powder, smear, liquid, etc.), its size, and, of course, on the type of experiment one wishes to perform (low temperature, high temperature, etc.).

C. Data Acquisition

The tasks of acquiring, storing and displaying the data can be performed in many ways. However certain basic procedures should be followed. For example, the signal from the microphone preamplifier should be processed by an amplifier tuned to the chopping frequency in order to maximize the signal-to-noise ratio. If phase as well as signal amplitude is desired, then a phase-sensitive lock-in amplifier should be used.

For a single-beam spectrometer, provision must generally be made to remove from the photoacoustic spectrum, any spectral structure resulting from the lamp, monochromator, and optics of the system. This normalization can be done conveniently by digitizing the analog signal from the tuned amplifier, and then performing a point-by-point normalization (i.e., division) with either a power meter reading or a previously recorded photoacoustic spectrum obtained with a black absorber. The equivalence of these two methods is illustrated in Fig. 8.

In a double-beam spectrometer, normalization can be performed in analog real-time fashion, by dividing the analog output from the tuned amplifier processing the sample signal, with the output derived from a reference signal. This reference output may be from a power meter or from a second photoacoustic cell.

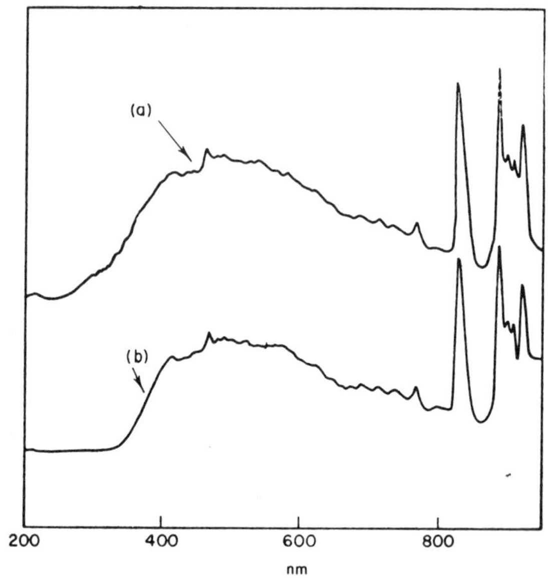

Fig. 8. (a) The photoacoustic spectrum of carbon black. (b) The power spectrum of the Xe lamp using an Si diode power meter.

With regard to the storage and display of the data, there are of course many possible schemes ranging from the relatively inexpensive chart recorder, to the sophisticated minicomputer.

IV. Photoacoustic Spectroscopy in Physics and Chemistry

In spite of the fact that photoacoustic spectroscopy of solids has been actively pursued at this time for only the past three years, it has already shown itself to be a very useful research and analytical tool in the fields of physics and chemistry. In general the applications of PAS in these fields can be divided into three main areas—bulk studies, surface studies, and deexcitation studies.

A. Bulk Studies

When the surface of a solid material is not highly reflective, photoacoustic spectroscopy will provide optical data about the bulk material itself. The PAS technique thus can be used to study insulator, semiconductor, and even metallic systems that cannot be studied readily by conventional absorption or reflection techniques, for example, substances in the form of powders, or those that are amorphous, or for some reason difficult to prepare for reflection studies.

In the case of insulators, photoacoustic spectra give direct information about the optical absorption bands in the material. This is illustrated in Fig. 9. Spectrum (a) shows the normalized PAS spectrum of some Cr_2O_3 powder in the region 200–1000 nm. Spectrum (b) is an optical absorption spectrum obtained by McClure (1963) on a 4-μm-thick Cr_2O_3 crystal. Spectrum (c) is a diffuse reflectance spectrum on Cr_2O_3 powder obtained

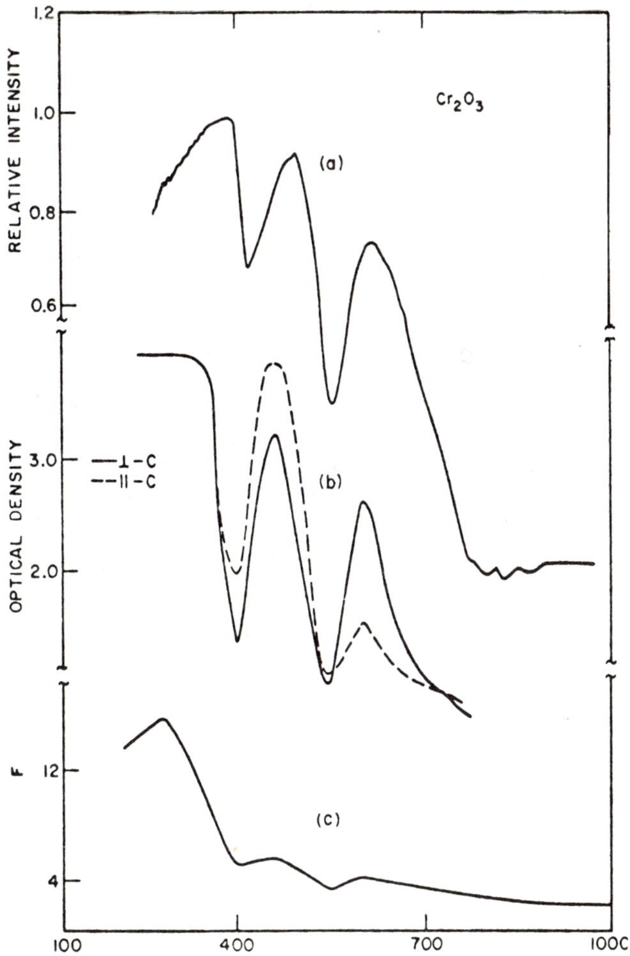

Fig. 9. (a) The normalized photoacoustic spectrum of Cr_2O_3 powder at 300°K. (b) The optical transmission spectrum of a 4.4 μ thick crystal of Cr_2O_3 at 300°K (McClure, 1963). (c) The diffuse reflectance spectrum of Cr_2O_3 powder at 300°K (Tandon and Gupta, 1970).

by Tandon and Gupta (1970). The two crystal field bands of the Cr^{3+} ion at 600 and 460 nm are almost as clearly resolved in the photoacoustic spectrum of the Cr_2O_3 powder as they are in the absorption spectrum of the Cr_2O_3 crystal, and they are much better resolved in the PAS spectrum than in the diffuse reflectance spectrum.

In Fig. 10a I show data obtained from an experiment on another inorganic insulator system, $CoMoO_4$. This experiment was performed in hope of obtaining further understanding of the hydrodesulfurization catalytic action of $CoMoO_4$ supported on alumina (Lipsch and Schmit, 1969). Both the high-temperature β-$CoMoO_4$ and low-temperature α-$CoMoO_4$ phases are available only as fine precipitates, and thus their optical spectra are not readily obtainable by conventional techniques. The photoacoustic spectra shown in Fig. 10a indicate that the β-$CoMoO_4$ has a charge-transfer band similar to that seen in the parent MoO_3, while the charge-transfer band in the α-$CoMoO_4$ has shifted noticeably to lower energy. In Fig. 10b we

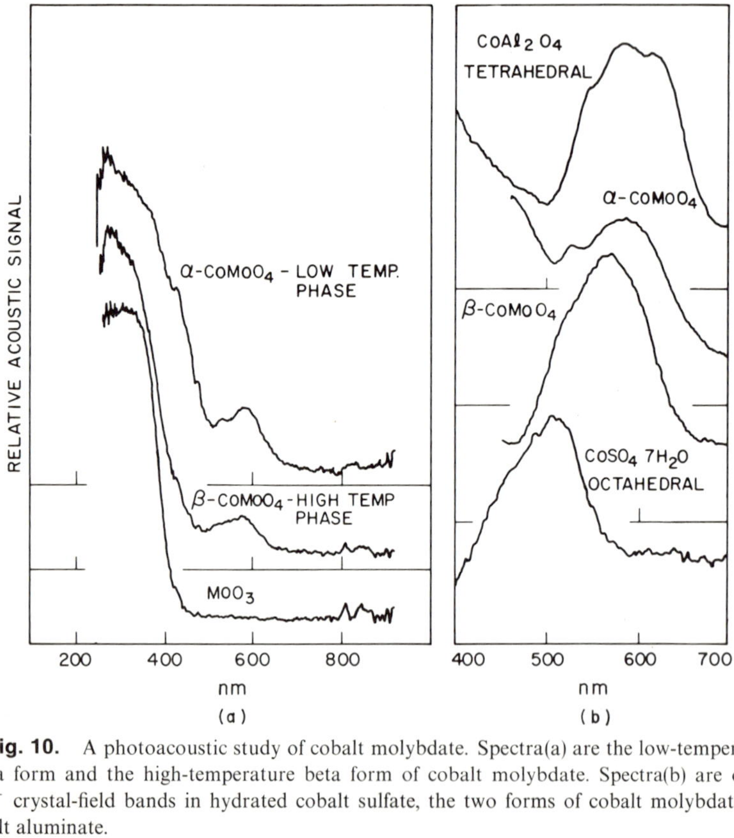

Fig. 10. A photoacoustic study of cobalt molybdate. Spectra(a) are the low-temperature alpha form and the high-temperature beta form of cobalt molybdate. Spectra(b) are of the Co^{2+} crystal-field bands in hydrated cobalt sulfate, the two forms of cobalt molybdate and cobalt aluminate.

have studied in more detail the crystal field bands of the Co^{2+} ions in $CoSO_4 \cdot H_2O$ (octahedral coordination) in β- and α-$CoMoO_4$, and in $CoAl_2O_4$ (tetrahedral coordination). These PAS spectra indicate that the Co^{2+} ions in both the β- and α-$CoMoO_4$ are in a distorted octahedral coordination, and that there is no significant change in the Co^{2+} d-electron configuration between the two $CoMoO_4$ phases. However, Fig. 10a indicates that the two phases of $CoMoO_4$ may have different catalytic properties because of the significant change in the charge-transfer band between the two phases. Similar studies on many other heretofore inaccessible catalytic systems may be performed with PAS.

In the case of semiconductors, both direct and indirect band transitions can be observed as shown in Figs. 11 and 12. In Fig. 11, results are shown

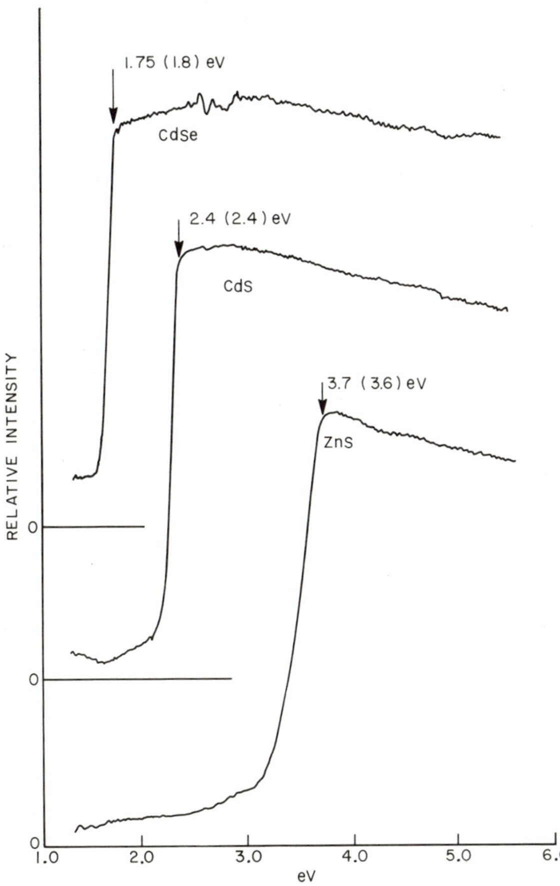

Fig. 11. Photoacoustic spectra of three direct-band semiconductors in powder form at 300°K.

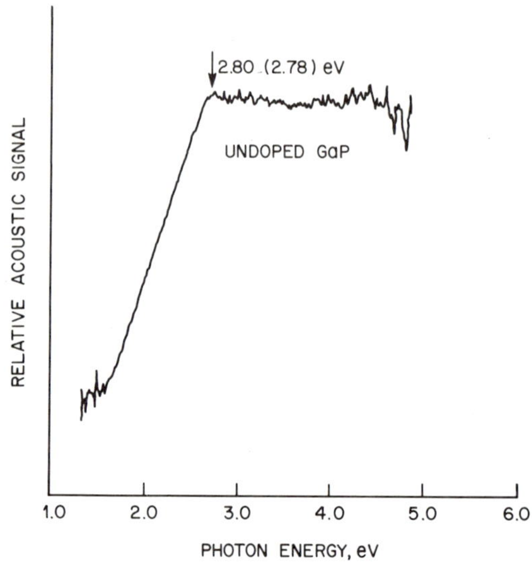

Fig. 12. Room-temperature photoacoustic spectrum of the indirect-band semiconductor GaP.

for three direct-band semiconductors all in powder form. The band edge as measured by the position of the knee in the PAS spectra agrees very well with the values recorded in the literature and given in parentheses (see, e.g., Pankove, 1971). Figure 12 shows the PAS spectrum of the indirect-band semiconductor GaP, also in the form of a coarse powder. Here again, the band edge value for the forbidden direct-band transition obtained from the PAS measurement agrees remarkably and rather surprisingly well with the literature value (Nelson *et al.*, 1964). The close agreement is surprising since one would expect a lower value for the knee due to opacity effects, arising from the strong indirect-band transitions in this semiconductor. It is these indirect-band transitions that give rise to the more gradual slope of the absorption edge seen in Fig. 12.

It is clear from these room-temperature experiments that PAS experiments performed at low temperature and higher resolution will allow one to study the excitonic and other fine structure in crystalline, powder, or amorphous semiconductors, that result from impurities, dopants, and other effects.

In the case of metallic systems, interband transitions and plasma excitations can also be studied by photoacoustic spectroscopy, provided that the reflection of light from the metal surface is not too great. This application is illustrated in Fig. 13, in which PAS spectra were taken on Cu and Ag metal powders. It is found that these spectra correspond to the reflection loss spectra of these metals, with the arrows in Fig. 13 indicating the positions

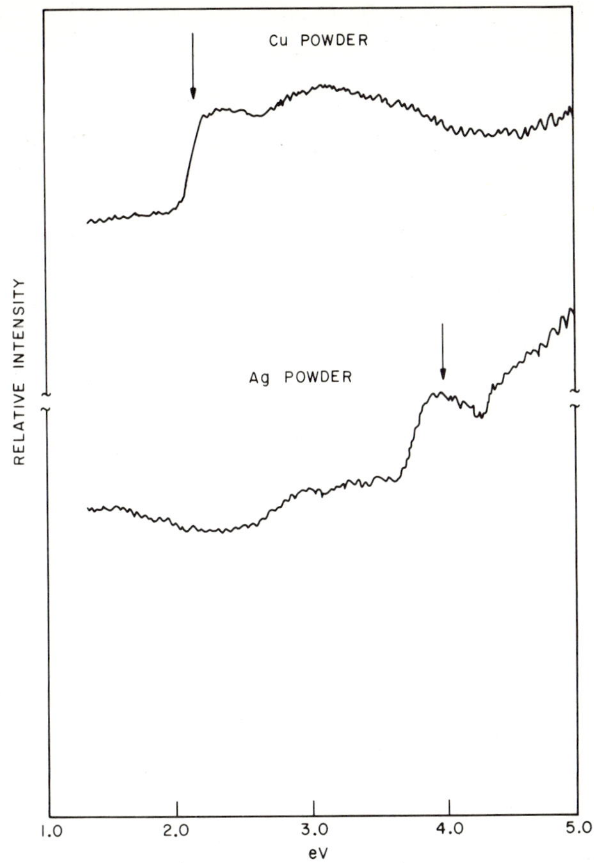

Fig. 13. Photoacoustic spectra of Cu and Ag powders. The arrows indicate positions of known structure in reflection loss spectra.

of major changes in the reflection losses of the metals, as measured by reflectivity techniques (Ehrenreich and Philipp, 1962).

A major advantage of PAS over conventional reflection spectroscopy in studies of semiconductors and metals is that highly reflecting surfaces are not needed (in fact, they are undesirable for bulk studies). This is particularly advantageous for studies on the new classes of organic and glassy semiconductor and metallic systems, which are often in powder or amorphous form. Furthermore, the need for ultraclean surfaces and thus the need for elaborate high-vacuum equipment so necessary for conventional reflection studies is not present for photoacoustic spectroscopy.

An example of results from a recent study on one-dimensional metallic systems (Rosencwaig *et al.*, 1976) is illustrated in Fig. 14, which shows

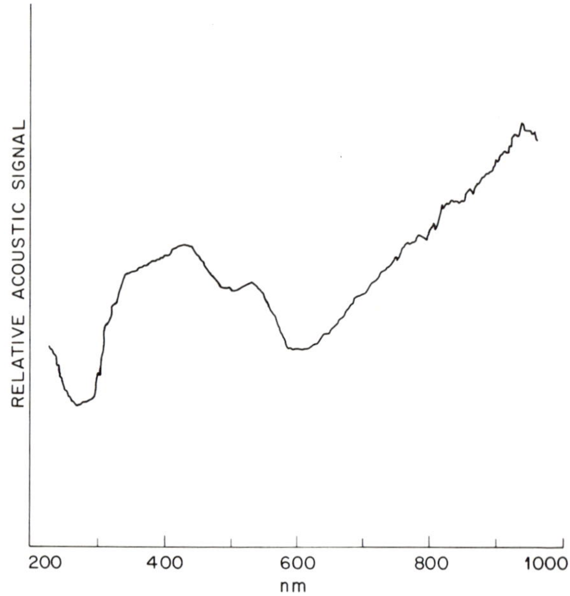

Fig. 14. Photoacoustic spectrum of a metalorganic one-dimensional conductor ($K_2Ir_2(CO)_4Cl_5$). This spectrum exhibits absorption band structure due both to localized and delocalized Ir ion electrons.

the PAS spectrum obtained on the linear chain one-dimensional conductor $K_2Ir_2(CO)_4Cl_5$. This compound like so many other one-dimensional conductors, is available at present only in the form of a fine powder and thus cannot be readily studied by conventional spectroscopic methods. The spectrum of Fig. 14 indicates that in $K_2Ir_2(CO)_4Cl_5$ the delocalized conduction band is a strong broad infrared band that drops off in the visible, and that the lower-wavelength region is dominated by localized Ir ion transitions. With photoacoustic spectroscopy one can therefore readily study the optical absorption spectra of these unusual compounds, and thus obtain new information about their electronic structures, a subject that is currently of considerable interest and controversy.

Another class of materials of considerable interest to both physicists and chemists are the organic liquid crystals. Because the solid, smectic, and nematic states of liquid crystals are highly light scattering, optical spectroscopy on these states cannot be readily performed. Nevertheless, optical data on these states may be very useful in providing information about the intermolecular interactions that play so important a role in determining the unusual properties of these materials.

In Fig. 15a we show data from a recent photoacoustic study (Rosencwaig *et al.*, 1975) of a class of liquid crystals. Since all members of this class are

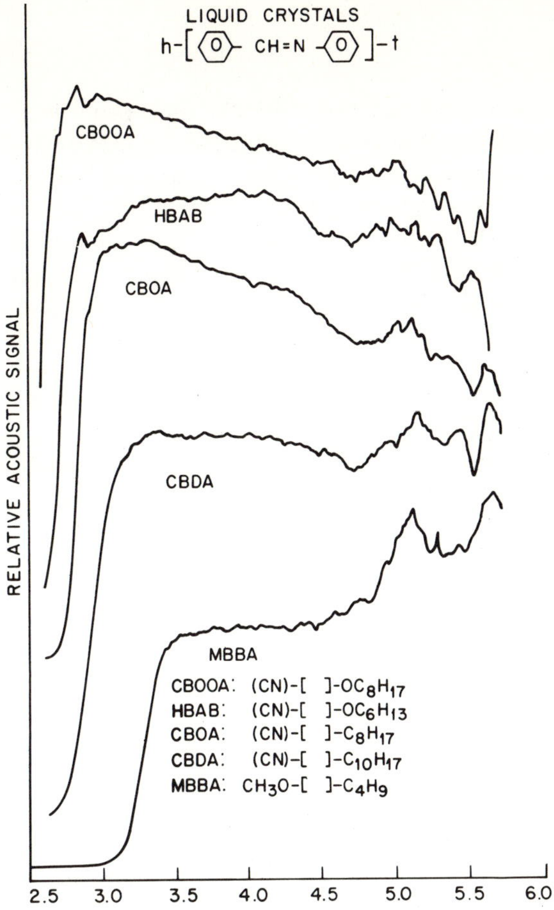

Fig. 15a

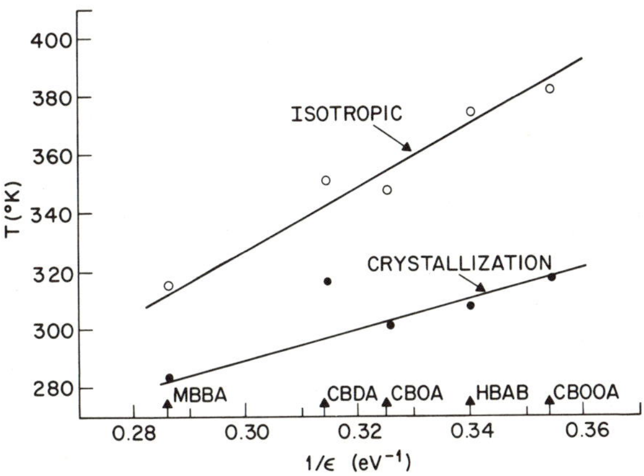

Fig. 15b

Fig. 15. A study of five liquid crystals. Their ultraviolet photoacoustic spectra are shown in (a). In (b) the temperatures at which these liquid crystals undergo their transitions to the isotropic form, and the crystal form, are plotted as functions of the reciprocal absorption-edge energies.

composed of molecules with essentially the same benzylidene aniline center, but with different heads and tails, the general features of the ultraviolet PAS spectra are quite similar. However there is a clear shift in the position of the primary absorption edge as we proceed from MBBA to CBOOA. This shift is likely a result of the different heads and tails, and thus is probably a measure of the different intermolecular interactions in these crystals. A correlation between the spectral data and the intermolecular interactions is clearly seen in Fig. 15b, where we have plotted the temperatures at which these materials crystallize and at which they go into their isotropic liquid state along the y axis, and the value ϵ^{-1}, where ϵ is the position of the primary absorption edge along the x axis. Thus, photoacoustic data on these liquid crystals can be used to obtain information about the intermolecular interactions in these crystals, and therefore enhance our understanding of the unusual properties of these compounds.

B. Surface Studies

Photoacoustic spectroscopy can be used to great advantage in the study of adsorbed and chemisorbed molecular species and compounds on the surface of metals, semiconductors, and even insulators. Such studies can be performed at any wavelength, provided that the substrate one is dealing with is either nonabsorbing or highly reflecting at this wavelength. Under either of these conditions, the PAS experiments will give the optical absorption spectra of the adsorbed or chemisorbed compounds.

Our first indication of the feasibility for using photoacoustic spectroscopy for surface studies came from an experiment performed with thin-layer chromatography (Rosencwaig and Hall, 1975). Thin-layer chromatography, or TLC, is a widely used and highly effective technique for the separation of mixtures into their constituent components. This technique is of considerable importance in the chemical, biological, and medical fields. The identification of the TLC-separated compounds directly on the TLC plates can be a fairly difficult procedure, however, particularly if reagent chemistry is inappropriate. Conventional spectroscopic techniques are unsuitable because of the opacity and light-scattering properties of the silica gel adsorbent on the TLC plates. Photoacoustic spectroscopy offers a simple and highly sensitive means for performing nondestructive compound identification directly on the TLC plates.

In Fig. 16 the PAS spectra taken on five different compounds that were separated and developed on TLC plates are shown on the left. These PAS spectra were taken directly on the plates themselves and were run in the

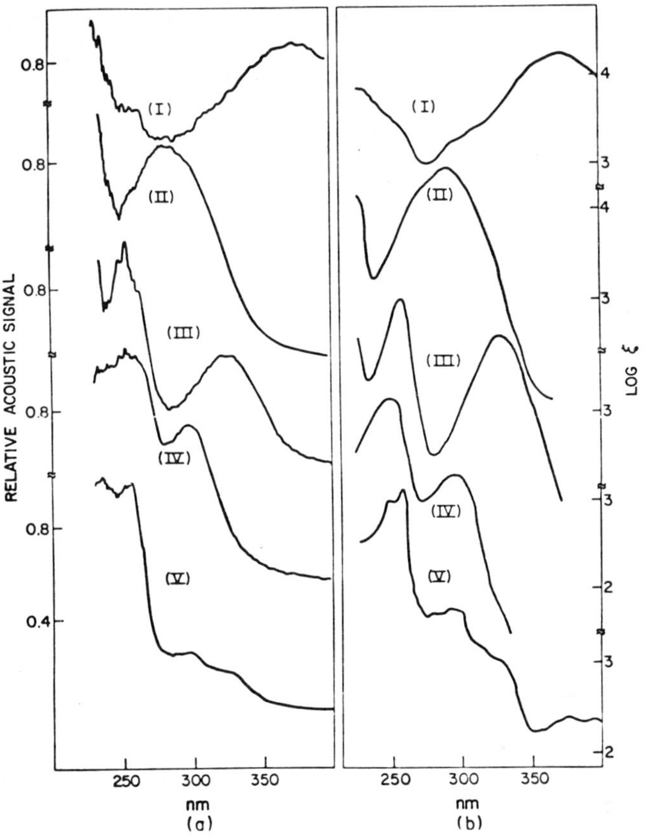

Fig. 16. A study on thin-layer chromatography. In (a) are shown the ultraviolet photoacoustic spectra of five compounds, *p*-nitroaniline (I), benzylidene acetone (II), salicylaldehyde (III), l-tetralone (IV), and fluorenone (V). These spectra were taken directly on the thin-layer chromatography plates. In (b) are shown the ultraviolet absorption spectra of the same five compounds in solution.

ultraviolet region of 200–400 nm. The five compounds starting from the top of Fig. 16 are (i) *p*-nitroaniline, (ii) benzylidene acetone, (iii) salicylaldehyde, (iv) l-tetralone, and (v) fluorenone. On the right we show, for the sake of comparison, the published ultraviolet absorption spectra of these compounds in solution (Perkampus *et al.*, 1966–1971). The strong similarity between the photoacoustic spectra and the optical absorption spectra permits a rapid and unambiguous identification of the compounds.

To test the sensitivity of the PAS technique in this application we performed the experiment shown in Fig. 17. Here we have the PAS spectra

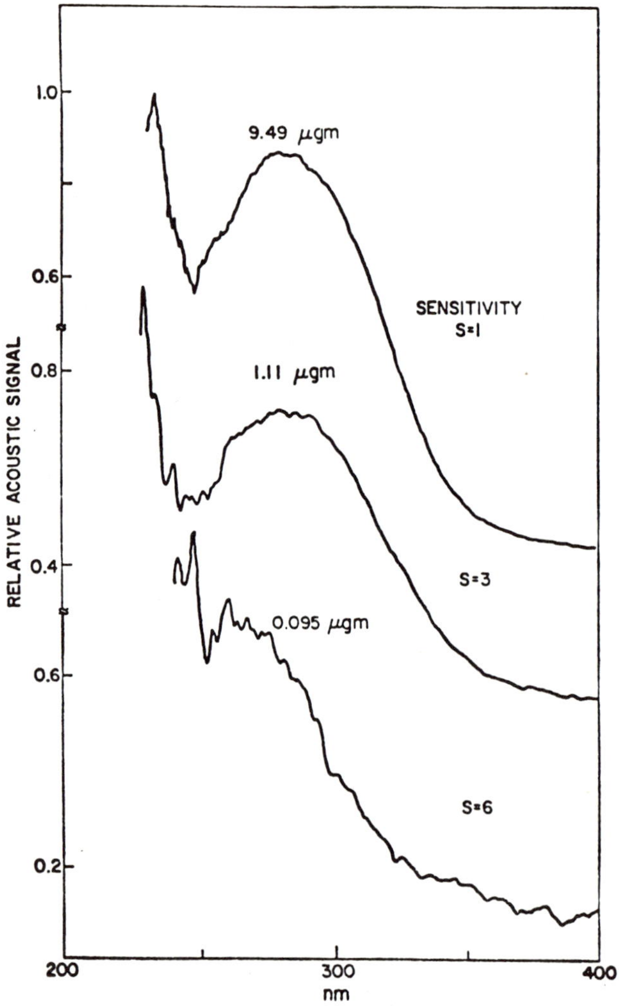

Fig. 17. Photoacoustic spectra on spots of benzylidene acetone on a TLC plate. The spots were developed from starting drops containing 9.49, 1.11, and 0.095 μgm of benzylidene acetone.

taken directly on a TLC plate on which benzylidene acetone spots of different concentrations were developed. The top spectrum was taken on a spot developed from a starting drop containing $\sim 10\ \mu gm$ of benzylidene acetone. The middle contained $\sim 1\ \mu gm$ and the bottom $\sim 0.1\ \mu gm$ in the starting drop. The spectra were taken on the developed spots, and these can be expected to have an even smaller amount of material than we started with. Nevertheless, we note that even for the case of $0.1\ \mu gm$, the main absorption band of the benzylidene acetone is visible. Knowing the amount of material present in this spot ($< 1\ \mu gm$), the size of the spot ($\sim 0.3\ cm^2$ in area), and the molecular weight of the compound (146), we estimate that we have roughly no more than one monolayer of the compound in this spot.

This experiment thus indicates the possibility that under certain conditions (low optical absorption or high reflectivity) of the substrate, the PAS technique may well be sensitive enough to detect and identify a monolayer of adsorbed or chemisorbed compound. Some new experiments (Rosencwaig and Schonhorn, 1975) on both metallic and nonmetallic surfaces have shown that monolayer detectability is achievable. The use of photoacoustic spectroscopy for such surface studies, particularly if used with high-resolution light sources such as tunable dye or infrared lasers, can lead to fundamental understandings of surface oxidation and reduction processes under a variety of conditions, and also to further knowledge about catalytic activity on solid surfaces. Other surface studies would include PAS studies of organic compounds and inorganic oxides purposely deposited on the surfaces of metals, semiconductors, and polymers for purposes of passivation. Such studies would yield data about the structure, valence, complexing, etc. of the deposited compound, information which is at present very difficult to obtain nondestructively.

An example of the use of PAS in surface passivation studies of metals (Rosencwaig and Schonhorn, 1975) is shown in Fig. 18. Spectrum (a) at the top is of a Cu surface that has been treated with benzotriazole, a known passivating agent. Spectrum (b) is of an identical but untreated Cu surface. Spectrum (a) differs markedly from spectrum (b) only in the wavelength region below 300 nm. Spectrum (c) is the difference spectrum produced by subtracting spectrum (b) from spectrum (a). Spectrum (c) thus represents the spectrum of the chemisorbed monolayer of benzotriazole. Spectrum (d) at the bottom is a PAS spectrum of some benzotriazole power. We note that spectrum (c) is quite different than spectrum (d), thus indicating that the chemisorbed benzotriazole has undergone significant structural change. We hope to establish, from such spectra, what changes have occurred in the benzotriazole upon chemisorption onto the Cu surface, and thus to understand how this compound passivates the surface of Cu metal.

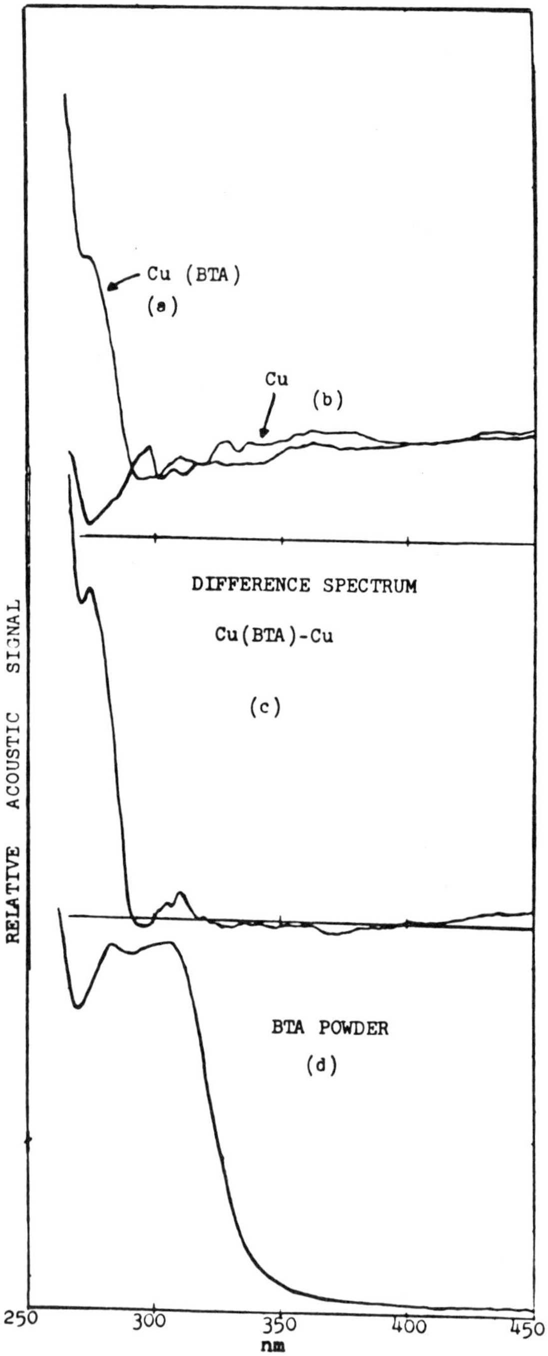

Fig. 18. A surface study. Spectrum (a) is a photoacoustic spectrum of a highly polished Cu surface treated with benzotriazole (BTA). Spectrum (b) is of an untreated surface. Spectrum (c) is the difference spectrum, showing the absorption bands of the monolayer of BTA. Spectrum (d) is a photoacoustic spectrum of bulk benzotriazole powder.

C. *Deexcitation Studies*

The photoacoustic effect measures the nonradiative deexcitation processes that occur in a system after it has been optically excited (or in more general terms, excited by any electromagnetic radiation, or in fact by any means whatsoever). This selective sensitivity of the PAS technique to the nonradiative deexcitation channel can be used to great advantage in the study of fluorescent (or phosphorescent) materials, and in the study of photosensitive substances.

1. *Fluorescent Studies*

When an optically excited energy level decays via fluorescence or phosphorescence, then little or no acoustic signal will be produced in the photoacoustic cell. This is illustrated in Fig. 19, where we consider the case of the fluorescent solid Ho_2O_3. Several of the trivalent rare earth ions, such as Ho^{3+}, have strongly fluorescent energy levels; that is levels that tend to deexcite through the emission of a photon, rather than through phonon or heat excitation. The upper PAS spectrum is of some Ho_2O_3 powder containing Co and F impurities. All of the lines present in this spectrum correspond to known Ho^{3+} energy levels, the positions of which are designated by the bars below. The dots indicate which of these levels are normally fluorescent. In this material the fluorescence is highly quenched by the presence of the Co and F impurities, and thus both the fluorescent and nonfluorescent lines appear in the PAS spectrum. The lower spectrum is of pure Ho_2O_3. Here all of the fluorescent levels have a greatly diminished relative intensity, since these levels are now deexciting through the emission of a photon rather than through heating of the solid.

In the case of fluorescent and phosphorescent materials, a combination of conventional fluorescence spectroscopy and photoacoustic spectroscopy will therefore provide data about both the radiative and nonradiative deexcitation processes within these solids. That is, the complete deexcitation process within these compounds can now be readily studied for the first time. By performing both fluorescence and photoacoustic spectroscopy as a function of temperature and compound composition, one can determine in a straightforward manner how these two variables affect the efficiencies and rates for the two deexcitation processes. Furthermore, since PAS spectroscopy gives phase as well as amplitude information, one can study exciton processes (random walk, lifetimes, etc.) in these materials as a function of temperature and dopant concentration.

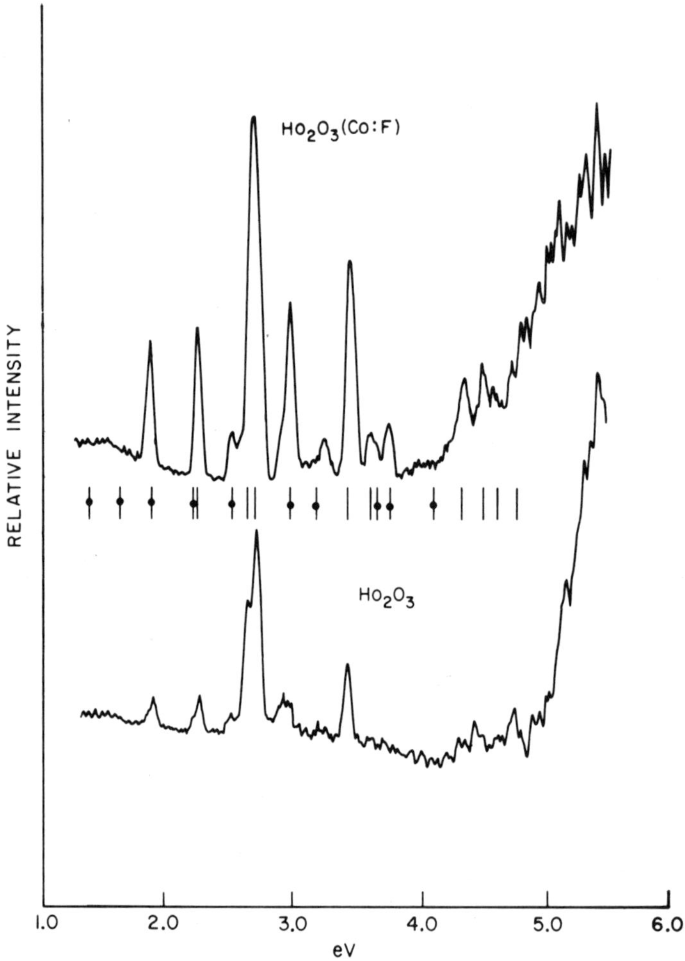

Fig. 19. Photoacoustic study of a fluorescent material. Cobalt and fluorine impurities quench the natural fluorescence of the Ho^{3+} ions in the holmium oxide here studied by photoacoustic spectroscopy. The upper spectrum is of doped, and the lower spectrum of undoped, holmium oxide. The fluorescent levels of Ho^{3+} are marked with dots.

2. *Photochemical Studies*

Another channel of deexcitation for absorbed light energy in some compounds is through photochemistry. Photoacoustic spectroscopy offers a unique tool for the study of photochemical process in solids. An illustration of this is shown in Fig. 20, where we have performed PAS experiments on the photosensitive material Cooper blue (2,2-dimethyl-4-phenyl-6-p-nitrophenyl-1, 3-diazabicyclo [3.1.0] hex-3-ene) (Rosencwaig and Trozollo,

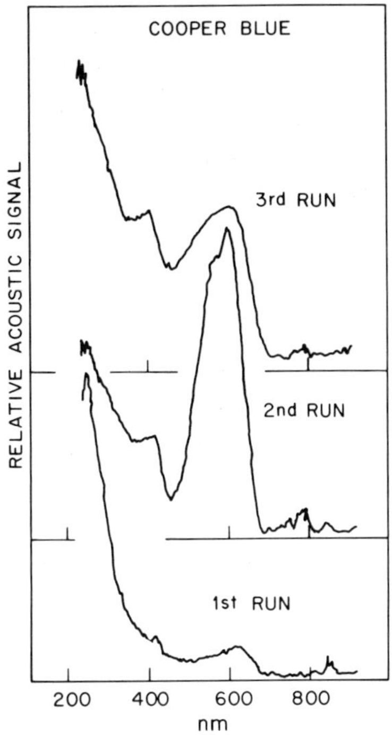

Fig. 20. Photoacoustic study of a photosensitive material. The Cooper blue was dark-adapted before the lower spectrum was made. The middle spectrum was run afterward, and then the upper spectrum, with only a few minutes separating the runs.

1975). This compound is colorless in the dark but turns a strong blue when exposed to light of short wavelength. The bottom spectrum of Fig. 20 is the PAS spectrum of dark-adapted Cooper blue. There is substantial ultraviolet but little visible absorption. The middle spectrum was obtained immediately after the first, and is quite different, showing two strong absorption bands in the visible spectrum. These are the bands that give Cooper blue its blue color. These bands arise from a photochemical change in Cooper blue wherein some of the photons absorbed in the short wavelength region have been utilized to break a ring in the Cooper blue molecule and thus create a new compound. The upper spectrum, run immediately after the middle spectrum, shows yet further changes, reflecting further photochemical and even photoinduced thermochemical processes.

Not only can one readily see the effects of photochemistry by means of photoacoustic spectroscopy, but one can also establish the activation spectrum for the photochemical process directly, by simply comparing the PAS spectrum with a conventional absorption spectrum. Information about the activation spectrum of photosensitive materials is at present quite difficult to obtain by other means. In addition, one can obtain, from the phase measurements of the photoacoustic signal, data about photochemical reaction rates, and even distinguish between true photochemical events and photoinduced thermochemical events. Photoacoustic studies on photosensitive materials will not only provide valuable basic information about the physical and chemical processes in these materials, but can also be of great benefit in the understanding of technologically important compounds such as photoresists, and in the study of photoinduced physical and chemical changes in polymers, plastics, and pigments.

V. Photoacoustic Spectroscopy in Biology and Medicine

Perhaps the greatest potentials for photoacoustic spectroscopy lie in the fields of biology and medicine, for here most of the materials to be studied are often in a form that makes them difficult if not impossible to study by any other optical technique. Although many biological materials occur naturally in a soluble state, many others are membrane bound or part of insoluble bone or tissue structure. These materials are insoluble and function biologically within a more or less solid matrix. Optical data on these materials is usually difficult to obtain by conventional techniques, since they are generally not in a suitable state for conventional transmission spectroscopy, and if solubilized, are often significantly altered. Photoacoustic spectroscopy, through its capability of providing optical data on intact biological matter, even on matter that is optically opaque, holds great promise as both a research and diagnostic tool in biology and medicine.

A. Hemoproteins

To illustrate the capabilities of PAS in biology we consider an experiment on cytochrome-C (Rosencwaig, 1973b), with the results shown in Fig. 21. The hemoprotein cytochrome-C plays an essential role in cellular respiration and has been extensively studied (e.g., Margoliash and Schijter, 1966). Since it is readily soluble in water, its optical absorption spectrum is well known. Figure 21a shows the optical absorption spectrum of the oxidized and reduced forms of cytochrome-C in aqueous solution obtained with a conventional spectrophotometer. Figure 21b shows the PAS spectra of oxidized and reduced cytochrome-C in lyophilized or powder form.

One notes that the photoacoustic spectra are qualitatively very similar to the optical absorption spectra. In particular, all of the differences between

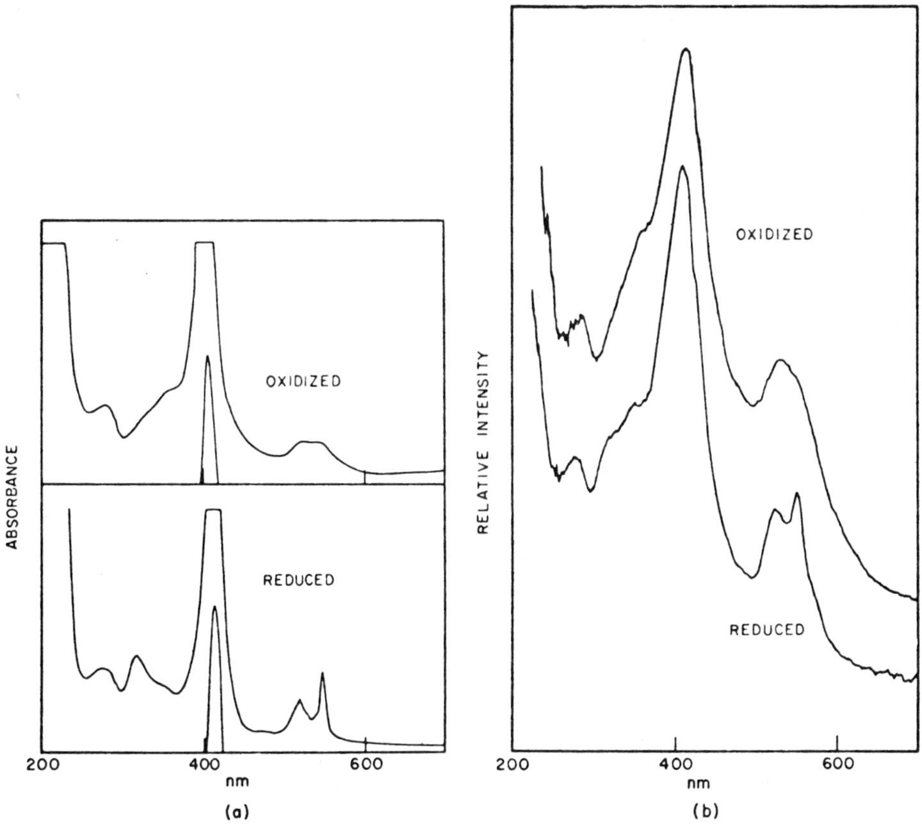

Fig. 21. A study of cytochrome-C. In (a) are shown the optical absorption spectra of oxidized and reduced cytochrome-C in aqueous solution. In (b) we have the photoacoustic spectra of oxidized and reduced cytochrome-C in solid form.

the oxidized and reduced forms visible in the absorption spectra, are also visible in the photoacoustic spectra. This experiment indicates that it is possible with photoacoustic spectroscopy to obtain spectra on biological material in solid form that are comparable to those obtained in solution. This capability becomes extremely important when the compound to be studied is insoluble.

The oxygen-carrying protein of red blood cells, hemoglobin (e.g., Antonini and Brunori, 1971) also displays a strong and characteristic heme absorption spectrum. Conventional absorption spectroscopy on whole blood, even when diluted in a suitable buffer, does not produce satisfactory results. The inadequacy of the conventional technique arises from the strong light-scattering properties of whole blood, owing primarily to the presence of the other protein and lipid material in the plasma and the red blood cells. In the

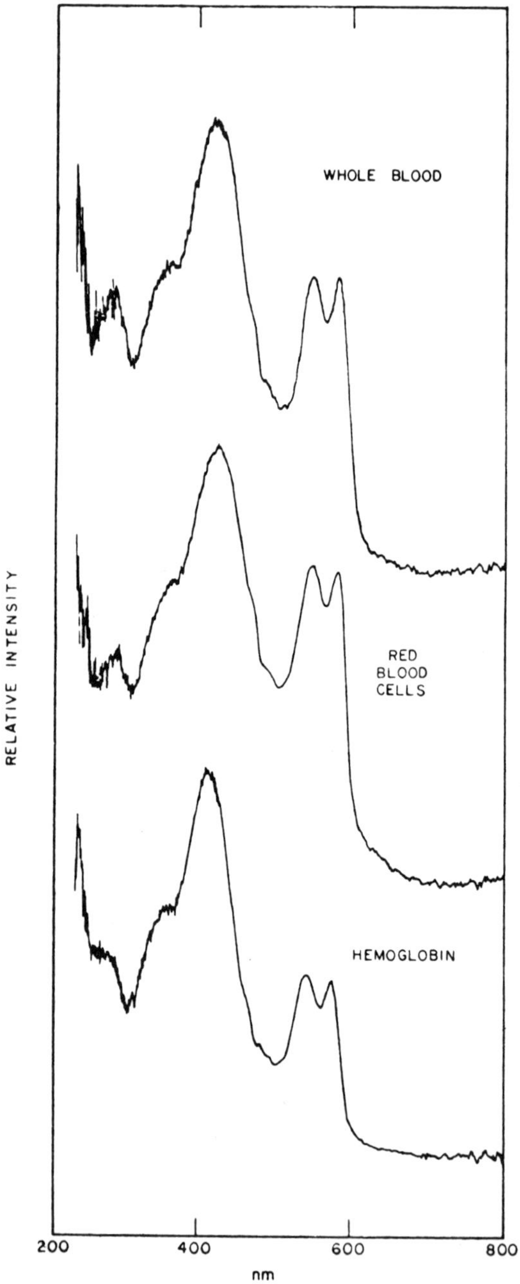

Fig. 22. Photoacoustic spectra of smears of whole blood, red blood cells, and hemoglobin.

conventional process, one first extracts the hemoglobin from the whole blood by centrifuge techniques and then obtains an absorption spectrum of an aqueous solution of the extracted hemoglobin. The PAS spectrum of a smear of whole blood in Fig. 22 exhibits the characteristic spectrum of oxyhemoglobin as clearly as in the PAS spectrum of the red blood cells and even of the extracted hemoglobin itself (Rosencwaig, 1973b). The presence of the other protein and lipid material in whole blood, which causes so much difficulty in conventional spectroscopy from light scattering, creates no problem in photoacoustic spectroscopy. Thus, it is now possible to study the hemoglobin directly in whole blood, that is, in situ, without resorting to extraction procedures.

B. Plant Matter

The unique capabilities of photoacoustic spectroscopy enable it to be used to obtain optical absorption data on even more complex biological systems such as green leaf and other plant matter. A PAS spectrum on an intact green leaf in Fig. 23 clearly shows all of the optical characteristics of

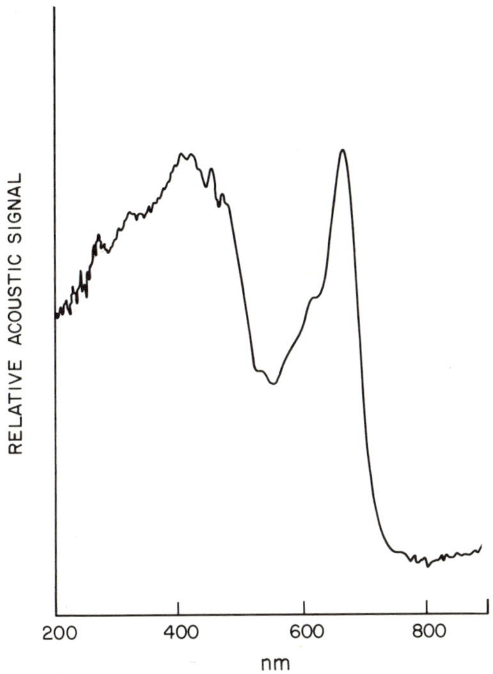

Fig. 23. A photoacoustic spectrum of an intact green leaf.

the chloroplasts in the leaf matter—the Soret band at 420 nm, the carotenoid band structure between 450 and 550 nm, and the chlorophyll band between 600 and 700 nm. Thus, PAS can be used to readily study intact plant matter, even living plant matter, and thus obtain valuable information about normal and abnormal plant processes and pathology.

Since one can study intact plant matter, it is clear that photoacoustic spectroscopy can be used as a quick and efficient screening tool in natural products chemistry (Rosencwaig and Hall, 1974). Using the PAS spectrum

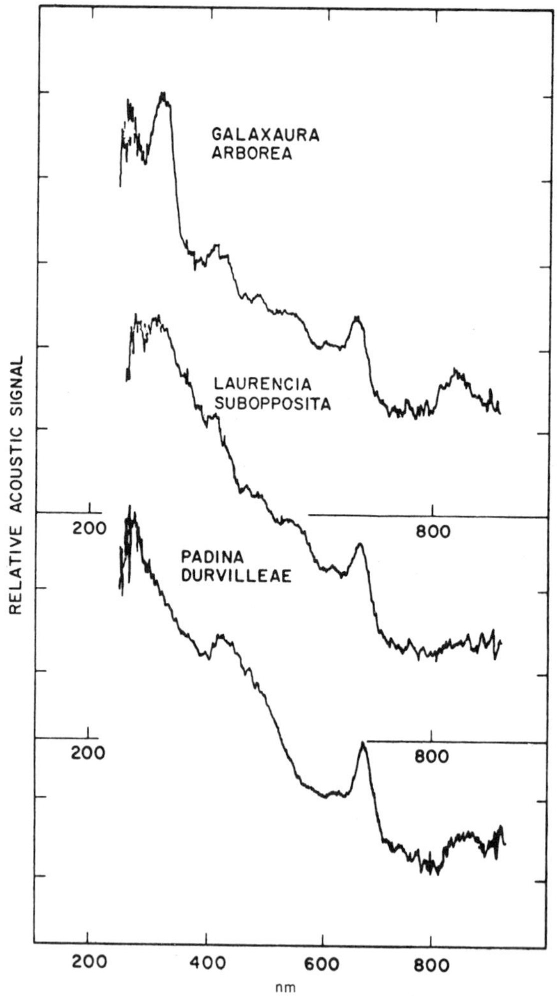

Fig. 24. Photoacoustic spectra obtained on a few milligrams of dried marine algae.

obtained from only a few milligrams of a natural source, such as a plant, animal, or microorganism, one can readily determine the types and relative concentrations of secondary metabolites present, that is, the chemical by-products of the organism's metabolism, compounds which may prove to have important physiological or biomedicinal value. At present such information can be arrived at only after laborious extraction and analysis procedures requiring many grams of the natural source.

The applications of PAS in natural products chemistry have been illustrated with a study on marine algae, a promising source of new and unusual natural products. (Scheuer, 1973). The photoacoustic spectra shown in Fig. 24 were taken on the algae after they had been air-dried, using less than 10 mg of material in each case. Although there is a considerable amount of spectral detail in all the PAS spectra, we were mainly interested in the region around 320 nm, a region where the sought-after aromatic or highly conjugated secondary metabolites could be expected to have an absorption band. The PAS spectra indicated that only two of the algae would yield any quantity of these aromatic compounds. Conventional extraction and analysis procedures have fully confirmed the photoacoustic data. It is clear therefore that photoacoustic spectroscopy can provide, in minutes, information about the compounds present in complex biological systems, and it is capable of doing this nondestructively, requiring only milligrams of material and no special sample preparation.

C. Tissue Studies

Photoacoustic spectroscopy can also play a potentially important role in medical studies. For example, we have shown that PAS can be used to monitor bacterial growth in cultures and on surfaces, and also that it may be possible to use the PAS spectra of bacteria to identify them. Another possible medical use is in the study of animal and human tissues, both hard tissues, such as teeth and bone, and soft tissues such, as skin, muscle, etc.

An example of a PAS experiment on soft tissue (Rosencwaig, 1975a; Rosencwaig and Lamola, 1975) is illustrated in Fig. 25, where photoacoustic spectra of guinea pig epidermis are shown. The top spectrum is of an epidermis that has been treated with a 2% in ethanol solution of tetrachlorosalicylanilide, or TCSA. This compound is known to be a highly effective antibacterial agent, but unfortunately also causes photosensitivity and other skin problems (Jenkins *et al.*, 1964). Why it has these side effects is not completely understood. The central spectrum is of control guinea pig epidermis. The bottom spectrum is the difference spectrum found by subtracting the control spectrum from the TCSA spectrum. The difference

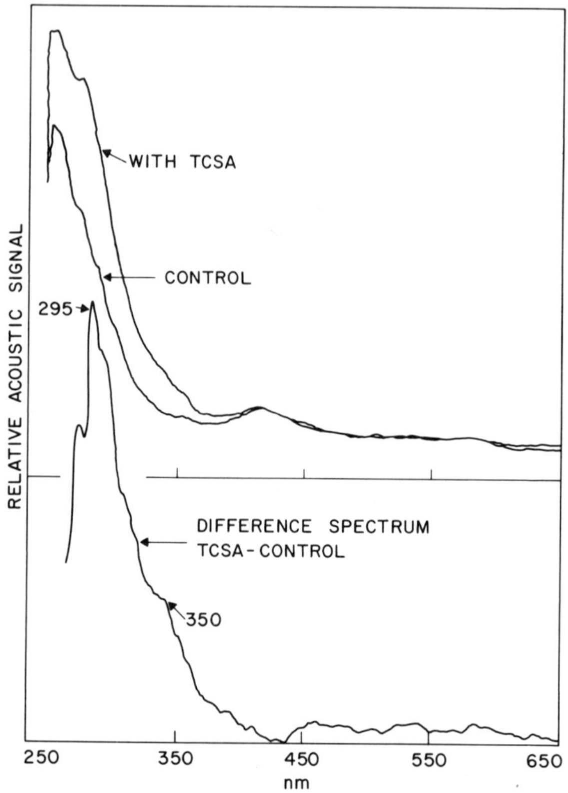

Fig. 25. A study of guinea pig epidermis. The upper spectrum is of a guinea pig epidermis which has been treated with tetrachlorosalicylanilide (TCSA). The middle spectrum is of untreated epidermis, and the bottom spectrum is the difference spectrum showing the absorption bands of the TCSA compound within the epidermis.

spectrum is thus the absorption spectrum of TCSA bound within the epidermis, the first such spectrum ever obtained by any technique. From this spectrum we can now establish the state of the TCSA compound in situ, that is, when it is incorporated into the skin, and thus learn more of its action on and within the skin, under various conditions. This example clearly illustrates the potential usefulness of photoacoustic spectroscopy in the study of natural and medical compounds present in human tissue.

In Fig. 26 we show the results of some PAS studies on new born rat stratum corneum (Rosencwaig and Pines, in press a, b). Conventional optical studies on intact stratum corneum, the uppermost layer of the skin, are difficult to pursue because of excessive light scattering and complications arising from the specimen's topology. We have applied the photoacoustic

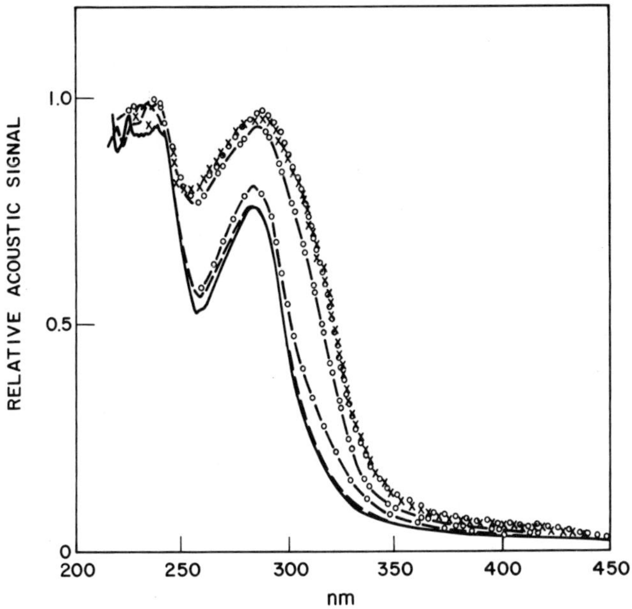

Fig. 26. A study of newborn rat stratum corneum: age after birth (——) $\frac{1}{4}$ hr; (---) 5 hr; (-●-) 10 hr; (-●●-) 21 hr; (●) 27 hr; (×) 60 hr. These ultraviolet photoacoustic spectra on intact newborn rat stratum corneum clearly exhibit marked changes in the protein absorption band resulting from initial maturation effects in the newborn rat stratum corneum.

technique to excised sections of intact new born rat stratum corneum, and as seen in Fig. 26 have obtained clean PAS spectra in the region of 220–450 nm.

The main absorption band visible in these spectra arises from the tryptophan and tyrosine amino acids present in the keratin protein of the stratum corneum (Fraser *et al.*, 1972). From studies of this spectrum we have been able to investigate the effects of hydration, temperature, chemical alteration and substantivity on the stratum corneum. As seen in Fig. 26 we have also observed that newborn rat stratum corneum shows marked spectral changes, line broadening as well as an increase in the tryptophan–tyrosine absorption during the first 30 hr after birth. We have found that these changes are attributable to a significant alteration in the α-keratin protein structure during this initial period of maturation.

Although these studies were performed on excised sections of skin, it is possible to obtain the photoacoustic spectrum of skin and other tissues in vivo. This can be done by using a suitable open-ended photoacoustic cell that is sealed by contact with the skin or tissue to be studied. The theory outlined in Section II clearly shows that for the proper choice of chopping frequency a photoacoustic spectrum of the tissue can be readily obtained, even though the entire sample (body) is opaque to transmitted light.

From an understanding of the PAS spectra of tissues we hope to obtain further information on the physical and biochemical properties of both normal and pathological tissue. As such information is accumulated and as correlations between spectral properties and pathologies emerge, the photoacoustic technique may prove to be a very valuable tool not only in the research but also in the diagnosis of tissue diseases such as cancer.

VI. Future Trends

Photoacoustic spectroscopy of solids is still in its formative stages, and its potential both as a research and analytical tool appears almost unbounded. Until the development of this technique, many materials, either natural or synthetic, could not be readily investigated by optical techniques, since these materials occur in the form of powders, or amorphous solids, or as semisolids such as smears, gels, oils, and so on. With photoacoustic spectroscopy, optical absorption data on virtually any solid or semisolid material can now be obtained.

In this chapter we have reviewed experiments with the photoacoustic effect in the fields of physics, chemistry, biology, and medicine. In all of these fields we have done no more than indicate some of the possible applications of this technique in these fields.

As photoacoustic spectroscopy of solids becomes better known, investigators in many different fields and with many different problems and orientations will adapt this technique to their own uses. Furthermore, since this is a spectroscopic technique that is not detector limited, it will surely be extended into the far ultraviolet and infrared regions of the optical spectrum, and quite possibly into other regions of the electromagnetic spectrum as well. The next few years should see an exciting period of growth for photoacoustic spectroscopy of solids.

References

Antonini, E., and Brunori, M. (1971). "Hemoglobin and Myoglobin in Their Reactions with Liquids." North-Holland Publ., Amsterdam.

Arnold, H. D., and Crandall, I. B. (1917). *Phys. Rev.* **10**, 22.

Bell, A. G. (1880). *Am. J. Sci.* **20**, 305.

Bell, A. G. (1881). *Philos. Mag.* **11**, 510.

Ehrenreich, H., and Philipp, H. R. (1962). *Phys. Rev.* **128**, 1622.

Fraser, R. D. B., MacRae, T. P., and Rogers, G. E. (1972). "Keratins, Their Composition, Structure and Biosynthesis," Thomas, Springfield, Illinois.

Jenkins, F. P., Welta, D., and Barnes, D. (1964). *Nature (London)* **201**, 827.
Lipsch, J. M. J. G., and Schmit, G. C. A. (1969). *J. Catal.* **15**, 163.
Margoliash, E., and Schijter, A. (1966). *Adv. Protein Chem.* **21**, 113.
* Maugh, T. H., II (1975). *Science* **188**, 38.
McClure, D. S. (1963). *J. Chem. Phys.* **38**, 2289.
Mercadier, M. E. (1881). *C. R. Acad. Sci.* **92**, 409.
Nelson, D. F., Johnson, L. F., and Gershenzon, M. (1964). *Phys. Rev.* **135**, A1399.
Pankove, J. I. (1971). "Optical Processes in Semiconductors," pp. 412–413. Prentice-Hall, Englewood Cliffs, New Jersey.
* Parker, J. G. (1973). *Appl. Opt.* **12**, 2974.
Perkampus, H. H., Sandeman, I., and Timons, C. J. (eds). (1966–1971). "DMS UV Atlas of Organic Compounds," Vol. I–V. Plenum Press, New York.
Preece, W. H. (1881). *Proc. R. Soc. London* **31**, 506.
Rayleigh Lord, (1881). *Nature (London)* **23**, 274.
Röntgen, W. C. (1881). *Philos. Mag.* **11**, 308.
Rosencwaig, A. (1973a). *Opt. Commun.* **7**, 305.
Rosencwaig, A. (1973b). *Science* **181**, 657.
Rosencwaig, A. (1975a). *Anal. Chem.* **47**, 592A.
Rosencwaig, A. (1975b). *Phys. Today* **28**, #9, 23.
Rosencwaig, A., and Gersho, A. (1975). *Science* **190**, 556.
Rosencwaig, A., and Gersho, A. (1976). *J. Appl. Phys.* **47**, 64.
Rosencwaig, A., and Hall, S. S. (1974). *Proc. Food-Drugs From the Sea Conf., 4th, Mayaguez, Puerto Rico, November.* Marine Technology Society, Washington (in press).
Rosencwaig, A., and Hall, S. S. (1975). *Anal. Chem.* **47**, 548.
Rosencwaig, A., and Lamola, A. A. (1975). Unpublished data.
Rosencwaig, A., and Pines, E. (in press a.) *J. Invest. Dermatol.*
Rosencwaig, A., and Pines, E. (in press b.) *Biochim. Biophys. Acta.*
Rosencwaig, A., and Schonhorn, H. (1975). Unpublished data.
Rosencwaig, A., Cladis, P. E., and Pasteur, G. A. (1975). Unpublished data.
Rosencwaig, A., Ginsberg, A. P., and Koepke, J. W. (1976). Inorg. Chem. **15**, 2540.
Scheuer, P. J. (1973). "Chemistry of Marine Natural Products." Academic Press, New York.
Tandon, S. P., and Gupta, J. P. (1970). *Phys. Status Solidi* **3**, 329.
Tyndall, J. (1881). *Proc. R. Soc. London* **31**, 307.
Viengerov, M. L. (1938). *Dokl. Akad. Nauk SSSR* **19**, 687.

Index

A
B 7
C 8
D 9
E 0
F 1
G 2
H 3
I 4
J 5